Plant Growth Regulators

TERTIARY LEVEL BIOLOGY

A series covering selected areas of biology at advanced under-graduate level. While designed specifically for course options at this level within universities and polytechnics, the series will be of great value to specialists and research workers in other fields who require a knowledge of the essentials of a subject.

Recent titles in the series:

Biology of Reptiles	Spellerberg
Biology of Fishes	Bone and Marshall
Mammal Ecology	Delany
Virology of Flowering Plants	Stevens
Evolutionary Principles	Calow
Saltmarsh Ecology	Long and Mason
Tropical Rain Forest Ecology	Mabberley
Avian Ecology	Perrins and Birkhead
The Lichen-Forming Fungi	Hawksworth and Hill
Plant Molecular Biology	Grierson and Covey
Social Behaviour in Mammals	Poole
Physiological Strategies in Avian Biology	Philips, Butler and Sharp
An Introduction to Coastal Ecology	Boaden and Seed
Microbial Energetics	Dawes
Molecule, Nerve and Embryo	Ribchester
Nitrogen Fixation in Plants	Dixon and Wheeler
Genetics of Microbes (2nd ed.)	Bainbridge
Seabird Ecology	Furness and Monaghan
The Biochemistry of Energy Utilization in Plants	Dennis
The Behavioural Ecology of Ants	Sudd and Franks
Anaerobic Bacteria	Holland, Knapp and Shoesmith
An Introduction to Marine Science (2nd edn.)	Meadows and Campbell
Seed Dormancy and Germination	Bradbeer

Plant Growth Regulators

JEREMY A. ROBERTS
Lecturer in Plant Physiology
Faculty of Agricultural Science
University of Nottingham

RICHARD HOOLEY
Senior Scientific Officer
Long Ashton Research Station
Bristol

Blackie

Glasgow and London

Published in the USA by
Chapman and Hall
New York

Blackie and Son Limited,
Bishopbriggs, Glasgow G64 2NZ
7 Leicester Place, London WC2H 7BP

Published in the USA by
Chapman and Hall
a division of Routledge, Chapman and Hall, Inc.
29 West 35th Street, New York, NY 10001–2291

British Library Cataloguing in Publication Data

Roberts, J.A. (Jeremy A.)
Plant growth regulators
1. Plants. Growth. Regulation
I. Title II. Hooley, R. III. Series
581.3′1

ISBN 0-216-92478-2
ISBN 0-216-92479-0 Pbk

Library of Congress Cataloging-in-Publication Data

Roberts, J.A. (Jeremy A.)
Plant growth regulators

(Tertiary level biology)
Bibliography: p.
Includes index
1. Plant regulators. I. Hooley, R. II. Title
III. Series
QK745.R62 1988   581.3′1   88-2864
ISBN 0-412-01661-3
ISBN 0-412-01671-0 (pbk.)

Printed in Great Britain by Bell and Bain Ltd., Glasgow

Preface

What are plant growth regulators? In the title, and throughout the text, we have adopted this expression to describe a population of endogenous molecules and synthetic compounds of similar structure that are believed to play important roles in the regulation of plant differentiation and development. For many years, plant scientists have endeavoured to understand the nature and action of plant growth regulators and, as a result, an awesome quantity of written material now exists describing these chemicals and their effects. In this book we have aimed to distil this wealth of information into a more digestible form, and in particular we have focused our attention on a critical appraisal of the literature.

The past few years have witnessed a change of emphasis in plant growth regulator research, which has been fuelled by powerful new techniques in molecular and cell biology. Today we can do more than just apply a plant growth regulator and quantify its effects; we have reached an exciting crossroads where plant scientists, molecular biologists and chemists can pool their expertise and apply it to the outstanding problems in this area. The combination of these three disciplines within the book is clear evidence of this.

In keeping with a volume of this size, we have assumed that the reader has a sound knowledge of plant physiology and biochemistry. However, wherever possible, we have highlighted useful reviews which provide background information, along with recent publications that have contributed significantly to the literature.

We are indebted to many of our colleagues for their invaluable advice and support throughout the writing of this book. In particular we wish to record our thanks to Professor Mike Black, Dr Jennie Elliott, Dr Nigel Given, Dr Peter Heddon, Dr Jane Taylor, Dr Mike Venis and Dr Mike Wright. Finally, one of us (JR) would like to apologize to Jennie and Sam for spending so many of his waking hours over the last year in front of a word processor.

JAR
RH

Abbreviations

ABA	abscisic acid
ACC	1-aminocyclopropane-1-carboxylic acid
Ade	adenine
AOA	aminooxyacetic acid
AVG	aminoethoxyvinylglycine
CCC	chlormequat
CK	cytokinin
CPP	copalylpyrophosphate
2,4D	2,4-dichlorophenoxyacetic acid
DPA	dihydrophaseic acid
EDTA	ethylenediaminetetra-acetic acid
EFE	ethylene-forming enzyme
EGTA	ethylene glycol-*bis*(β-aminoethylether)-tetra-acetic acid
ELISA	enzyme-linked immunoassay
FC	fusicoccin
GA	gibberellin
GC	gas chromatography
GCMS	combined gas chromatography–mass spectrometry
GGPP	geranylgeranylpyrophosphate
HPLC	high performance liquid chromatography
IAA	indole-3-acetic acid
IBA	indole-3-butyric acid
LDP	long-day plant
MACC	malonyl ACC
MTA	methylthioadenosine
NAA	naphthaleneacetic acid
NPA	naphthylphthalamic acid
ODA	2,7-dimethyl-2, 4-octadienedioic acid
PA	phaseic acid
PG	polygalacturonase
PGR	plant growth regulator
PP333	paclobutrazol
RIA	radioimmunoassay

SAM	S-adenosylmethionine
SDP	short-day plant
SIM	selective ion monitoring
STS	silver thiosulphate
TIBA	triiodobenzoic acid

Contents

Preface   v
Abbreviations   vïi

CHAPTER 1   Introduction — the challenge of PGR research   1

CHAPTER 2   Biosynthesis and metabolism — more than making and breaking   4

    2.1   Auxins   4
       2.1.1 Background and biosynthesis   4
       2.1.2 Metabolism   7
       2.1.3 Biosynthesis of IAA in *Agrobacterium tumefaciens* tumours   8
    2.2   Gibberellins   8
       2.2.1 Background and biosynthesis   8
       2.2.2 Metabolism   15
    2.3   Cytokinins   15
       2.3.1 Background and biosynthesis   15
       2.3.2 Metabolism   20
    2.4   Abscisic acid   21
       2.4.1 Background and biosynthesis   21
       2.4.2 Metabolism   23
    2.5   Ethylene   26
       2.5.1 Background and biosynthesis   26
       2.5.2 Metabolism   29
    2.6   Polyamines   31
    2.7   More than making and breaking   32

CHAPTER 3   Extraction, identification and quantification — the state of the art   34

    3.1   Extraction   34
    3.2   Purification   35
    3.3   Identification and quantification   36
       3.3.1 Bioassays   36
       3.3.2 Chemical methods   38
       3.3.3 Immunological methods   38
       3.3.4 Gas chromatography   41
       3.3.5 Gas chromatography–mass spectrometry   42
       3.3.6 High performance liquid chromatography   44
    3.4   The state of the art   46

CHAPTER 4   Hormones and the concept of sensitivity — a rational approach   49

4.1   The hormone concept   49
      4.1.1 Sites and regulation of biosynthesis   50
      4.1.2 Transport and its regulation   55
      4.1.3 Target cells   61
4.2   The concept of sensitivity   62
4.3   Hormones and the concept of sensitivity — a rational approach   63

CHAPTER 5   Cellular differentiation and morphogenesis   68

5.1   Juvenility   68
5.2   Flowering   69
      5.2.1 Photoperiodism   69
      5.2.2 Vernalization   72
5.3   Sex expression   73
5.4   Vascular differentiation   74
5.5   Morphogenesis   77
5.6   Conclusions   79

CHAPTER 6   Seed development, dormancy and germination   80

6.1   Seed development   80
6.2   Dormancy   83
6.3   Germination   86
6.4   Mobilization of storage reserves   88
6.5   Conclusions   92

CHAPTER 7   Root and shoot development   93

7.1   Growth   93
7.2   Tropisms   100
      7.2.1 Gravitropism   100
      7.2.2 Phototropism   105
7.3   Apical dominance   106
7.4   Bud dormancy and tuberization   110
7.5   Conclusions   112

CHAPTER 8   Leaf, flower and fruit development   114

8.1   Growth   114
8.2   Regulation of stomatal aperture   117
8.3   Epinasty and hyponasty   119
8.4   Ripening   121
8.5   Senescence   125
8.6   Abscission   129
8.7   Conclusions   132

CHAPTER 9   Receptors — sites of perception
                   or deception?   134

        9.1   Binding studies   134
               9.1.1 Theory   134
               9.1.2 Practice   137
        9.2   Binding sites for PGRs   139
               9.2.1 Auxins   139
               9.2.2 Gibberellins   144
               9.2.3 Abscisic acid   145
               9.2.4 Cytokinins   146
               9.2.5 Ethylene   149
        9.3   Sites of perception or deception?   150

CHAPTER 10   Mechanisms of action — towards a molecular
                   understanding   151

        10.1   Regulation of ion movement   151
        10.2   Regulation of gene expression   153
               10.2.1 Auxins   155
               10.2.2 Gibberellins   155
               10.2.3 Abscisic acid   158
               10.2.4 Ethylene   158
        10.3   Second messengers   159
               10.3.1 Calcium   160
               10.3.2 Calmodulin   161
               10.3.3 cAMP   161
               10.3.4 Protein phosphorylation   161
               10.3.5 Inositol phospholipids   163
        10.4   Towards a molecular understanding   163

CHAPTER 11   Commercial applications for PGRs — thought
                   for food?   164

        11.1   Auxins and related compounds   165
        11.2   Gibberellins and growth retardants   166
        11.3   Cytokinins and related compounds   168
        11.4   Abscisic acid and anti-transpirants   169
        11.5   Ethylene-generating or suppressing compounds   170
        11.6   Thought for food — food for thought   173

                   References and further reading   175
                   Index   185

CHAPTER ONE

INTRODUCTION — THE CHALLENGE
OF PGR RESEARCH

An understanding of how plants grow is not only of intrinsic interest
to the fundamental scientist but is also of far-reaching consequence to
the applied biologist. As a result, there has been a wealth of research
undertaken over the last fifty years to probe the mechanisms responsible for regulating plant differentiation and development. A common
message that has emerged from many of these studies is that the time-
course of developmental programmes is regulated by specific endogenous chemicals. These compounds were initially christened 'plant
hormones', but more recently this term has fallen from favour, and
both 'plant growth substances' and 'plant growth regulators' have been
adopted into modern usage. Although it can be readily argued that each
of these three descriptions is inappropriate, for the sake of consistency
the term plant growth regulator (PGR) has been used throughout this
volume.

There are five classes of compounds in the premier division of endogenous PGRs. These are auxins, gibberellins (GAs), cytokinins (CKs),
abscisic acid (ABA) and ethylene. Broadly speaking, the auxins and
GAs have been classified as regulators of cell elongation, the CKs as
regulators of cell division, ABA as a general inhibiting influence, and
ethylene as a volatile with a finger in numerous 'developmental pies'.
Descriptions such as these are clearly simplistic, and represent vain
attempts to dispense PGRs into neat compartments on the basis of their
effects on plant tissues. This common practice is misguided, not only
because it implies that a PGR has a similar impact on the behaviour of
any plant cell, but also, and more significantly, because it assumes that
the application of a PGR must mimic its effects *in situ*. The balance
of evidence indicates that neither of these assumptions is correct. In
addition, it is naïve to believe that the five groups of PGRs described
above may be the only molecules that can influence plant growth and
development; an increasing number of other compounds such as the
polyamines and the brassinosteroids are being reported to have potent
effects on plant tissues.

No book of this size can realistically lead the reader step by step from the discovery of the various PGRs to the forefront of modern PGR research. Therefore a certain amount of background knowledge has had to be assumed. Furthermore, it has not been our intention to produce an encyclopaedic work describing every developmental event where PGRs have been implicated. Rather we have preferred to concentrate on a critical appraisal of PGRs in the regulation of representative processes.

Like many other contributors to this field, we begin by considering how the major groups of PGRs are synthesized and metabolized by plant tissues. However, the emphasis of this chapter is not on the reader to recall the structure of every putative biosynthetic intermediate or metabolite that has ever been isolated, but to appreciate that it is a knowledge of the regulatory mechanisms responsible for the making and breaking of PGRs that will ultimately lead us to the position where we can manipulate the endogenous levels of these compounds at will. Gross changes in the concentrations of PGRs in certain tissues clearly can occur during the ontogeny of a plant; however, many critical changes may be much more subtle than this and involve a change in intracellular compartmentalization. There is thus a need for highly sensitive and specific techniques to be developed which can be used to quantify and localize PGRs in individual cells, and the rapid progress which is being made in this area is reviewed in Chapter 3. The succeeding chapter enters the highly charged debate of whether development may be regulated by the concentration of a PGR or the sensitivity of a cell to that PGR. This controversy has been instrumental in a critical reappraisal of the role of PGRs in differentiation and development and this is the main theme of Chapters 5 to 8. Chapters 9 and 10 consider the dramatic impact that plant molecular biology has had on PGR research. For instance, with the adoption of this approach it has been unequivocally demonstrated that the expression of a spectrum of genes is influenced by PGRs, and in some cases the identity of the proteins which they encode is known. Furthermore, the identification of the regulatory DNA sequences and other elements responsible for placing a gene under the influence of a PGR are clearly imminent. At the other end of the chain of cellular events, the identification and purification of PGR receptors is awaited with eagerness, while the search for second messengers is being carried out with renewed enthusiasm. Last, and by no means least, in the final chapter attention is turned to the use of PGRs and their analogues as commercial regulators of plant growth and development. Many compounds that have been 'discovered' by industrial screening can be shown to act either directly or indirectly through their capacity to modulate PGR levels or effects. As our knowledge of the mode of action of PGRs improves, so too will our ability to target specific physiological and biochemical processes for manipulation to the benefit of the agricultural and horticultural industries.

It is our view that PGR research, fuelled by the powerful techniques of molecular and cellular biology, has entered a stimulating and purposeful new phase. For both the novice and experienced worker in this field, there are exciting challenges ahead.

CHAPTER TWO

BIOSYNTHESIS AND METABOLISM — MORE THAN MAKING AND BREAKING

To set the scene for this book it is important for the reader to have some knowledge of the history of the discovery of the major PGRs. With this in mind, we have therefore included a brief résumé of the background behind their isolation and characterization; however, the main thrust of this chapter is centred on the mechanisms by which PGRs are built and broken down. In particular, it is our intention to emphasize the importance of biosynthesis and metabolism in the regulation of endogenous PGR levels.

2.1 Auxins

2.1.1 Background and biosynthesis

The sequence of events leading to the isolation and characterization of the major auxin in higher plants, indole-3-acetic acid (IAA), can be traced back to the experiments carried out by Charles and Francis Darwin in 1880. These workers found that if the tip of a coleoptile was shaded with a small cap, then unilateral light would not induce a phototropic response. Since unilateral irradiation stimulated bending throughout the length of unshaded coleoptiles, the Darwins concluded that the coleoptile tip perceived the light and transmitted some 'influence' to the other tissues which induced differential growth. Some years later, Boysen-Jensen (1911) demonstrated that the 'influence' could pass through gelatin, and concluded that it was probably chemical in nature.

Attempts to extract the chemical into water proved unsuccessful; however, by modifying Boysen-Jensen's original procedure, Fritz Went (1928) succeeded in collecting diffusates from excised coleoptiles into agar blocks which, when placed asymmetrically on freshly decapitated coleoptiles, induced bending. This significant breakthrough not only confirmed the existence of a growth-promoting compound in coleoptile tips, but also resulted in the development of a quantitative bioassay for auxins. It was a short step from here to the identification of IAA as a compound with 'auxin-like' activity (Kogl *et al.*, 1934), and later the demonstration that this was a primary auxin of higher plant tissues (Haagen-Smit *et al.*, 1946). Since then, there has been an intensive search for other naturally occurring auxins, and both 4-chloro-3-indoleacetic acid and phenylacetic acid have been identified in this category. Although 4-chloro-3-indoleacetic acid is more active than IAA in certain auxin bioassays, it may have limited significance as an endogenous PGR since it has only been detected in immature seeds of *Pisum sativum* and *Vicia faba*. In contrast, phenylacetic acid exhibits lower activity than IAA in the standard bioassays, but substantial amounts have been found in tobacco and tomato tissues, and therefore it is conceivable that it plays a significant role as an auxin in some species.

The close similarity in chemical structure between tryptophan and IAA makes the ubiquitous amino acid a logical candidate as the precursor of this auxin. However, although a number of different pathways have been proposed to account for the biosynthesis of IAA from tryptophan (see Figure 2.1), definitive evidence in support of any one of them has yet to be obtained. The most favoured pathway involves the transamination of tryptophan to indole-3-pyruvic acid (see Cohen and Bialek, 1984). Although unstable, this α-keto acid has been detected in a variety of plant tissues, and tryptophan aminotransferase activity has been reported in *Nicotiana tabacum* callus tissue. It has been proposed that the indole-3-pyruvic acid, once formed, is decarboxylated to indole-3-acetaldehyde which is finally converted to IAA. The enzymes necessary for these latter reactions have been found in cell-free extracts from plant tissues. Other biosynthetic routes of IAA from tryptophan have also been postulated. One possibility is that the amino acid is decarboxylated to tryptamine before being converted to indole-3-acetaldehyde and finally to IAA. It has also been suggested that indole-3-acetaldehyde is an intermediate in the indole-3-acetaldoxime pathway which is a series of biosynthetic reactions thought to be characteristic of the *Brassica* family. An alternative intermediate in this pathway would be indole-3-acetonitrile. The message from many of the studies on auxin biosynthesis, is that tryptophan *can* be converted to IAA by plant tissues. However, to what extent this reaction occurs *in situ* is open to question, since the complexities of studying the process are compounded by trivial conversion during chemical extraction, conversion arising out of microbial contamination, and conversion by damaged cells during the feeding process. Until *in-vitro* cell-free biosynthetic systems

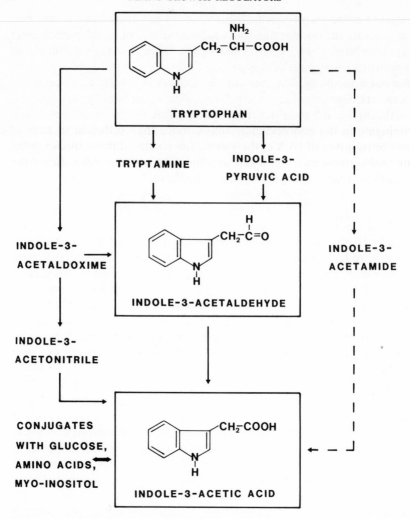

Figure 2.1 Biosynthetic pathways of IAA. Pathway denoted by the dotted line has been proposed to occur in cells infected by the bacterium *Agrobacterium tumefaciens*. (Scheme modified from Weiler and Schroder, 1987.)

are routinely successful, or IAA biosynthetic mutants are identified and fully characterized, the evidence in favour of the conversion of trypto-phan to IAA may remain equivocal.

An alternative process by which auxin levels in plant tissues could be elevated is the release of IAA from amide-linked or ester conjugates. In general, the former compounds have been found to predominate in legume seedlings and the latter in seedlings of cereals. Evidence to support the conjugate hydrolysis hypothesis has come from studies on *Zea mays* seedlings. For instance, if ^{14}C-IAA-myo-inositol is applied to maize endosperm tissue, ^{14}C-IAA can be detected in the growing shoot. Furthermore, it has been shown that the rate of disappearance of the conjugate in the endosperm *in vivo* is more than sufficient to account for the turnover of IAA in the shoot. The results of these studies indicate that conjugates could act as an important source of IAA during the growth and development of young seedlings.

2.1.2 Metabolism

The preceding discussion has highlighted the potential significance of conjugates in regulating levels of IAA. However, the process of metabolism of PGRs is of general importance in determining the endogenous levels of these compounds. Studies on PGR metabolism have relied primarily on three approaches. Radiolabelled PGRs have been fed to tissues and the resultant metabolites identified; potential metabolites have been extracted from plants and their concentrations quantified; and the properties of putative metabolizing enzymes have been studied.

The first enzyme capable of metabolizing IAA was characterized from pea seedlings by Tang and Bonner in 1947. This enzyme catalysed IAA oxidation. Peroxidases with this property have since been discovered in a spectrum of plant species, and these enzymes have been shown to catalyse the decarboxylation of IAA to the main products, 3-methyleneoxindole or indole-3-aldehyde, *in vitro*. Identification of these putative metabolites as endogenous constituents of plant tissues has been rare; however, recently indole-3-methanol has been identified in etiolated *Pinus sylvestris* seedlings. This compound has been proposed to be the first product of the *in-vitro* metabolism of IAA by horseradish peroxidase, and it has been suggested that it is subsequently converted to indole-3-aldehyde. Indole-3-methanol has also been shown to be the major metabolite produced after feeding ^{14}C-IAA to a chloroplast suspension from pea seedlings. The rate of decarboxylation of IAA in this system was significantly enhanced by light. Indole-3-methanol has yet to be confirmed as an endogenous constituent of pea chloroplast fractions; however, this study may assist in the identification of the *in-vivo* pathway of IAA decarboxylation and the cellular locations where it takes place.

Oxindole-3-acetic acid was first proposed as a natural metabolite of IAA by Klambt in 1959. However, only recently has it been identified as an endogenous constituent of *Zea mays* endosperm and shoot tissue. Feeding experiments using *Zea mays* seedlings have shown that over 60% of the ^{14}C-IAA applied to the root or shoot tissues is metabolized within 2 h of treatment. During this period negligible decarboxylation of IAA can be detected. An analysis of the ^{14}C-products extracted from the tissues has revealed that the most prominent one co-chromatographs with authentic oxindole-3-acetic acid. The conclusion from this study, is that at least two pathways of IAA metabolism must exist in plant tissues (see Bandurski, 1984). Whether peroxidative decarboxylation, or oxidation to oxindole-3-acetic acid (see Figure 2.2), represents the major pathway of IAA catabolism remains unknown. The answer could vary according to species and even plant tissue.

2.1.3 Biosynthesis of IAA in Agrobacterium tumefaciens tumours

Agrobacterium tumefaciens is a bacterium which induces 'crown gall' disease in many dicotyledonous species. Infection is associated with the development of neoplastic growth, and these tumours contain high concentrations of IAA. It has been demonstrated that, upon infection, *Agrobacterium* transfers a portion of its DNA, the T-DNA, into a plant cell, and this becomes integrated into the host's genome. Two of the genes transferred encode enzymes that together produce IAA (Thomashow *et al.*, 1986). The first encodes tryptophan monooxygenase which has the ability to convert tryptophan to indoleacetamide. The second gene codes for indoleacetamide hydrolase which converts indoleacetamide to IAA. The discovery of these genes has made it possible to engineer plants that produce high levels of IAA at tissue-specific sites. Clearly, such plants will make an important contribution to our understanding of the role of IAA in plant differentiation and development.

2.2 Gibberellins

2.2.1 Background and biosynthesis

The discovery of the GAs is a fascinating story which commences with the studies carried out by Japanese plant pathologists on the 'bakanae'

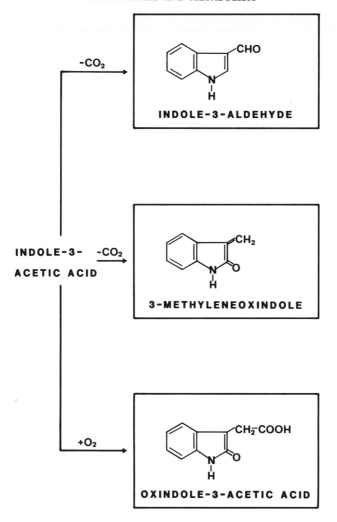

Figure 2.2 Structure of some of the products of metabolism of IAA.

disease of rice. 'Bakanae' literally means 'foolish seedling' in Japanese, which was considered to be an apt description of a disease which made infected rice plants grow so tall that they fell over. In 1926, Kurosawa demonstrated that a sterile culture filtrate from the fungus known to be responsible for causing the disease, *Gibberella fujikuroi*, could induce abnormal growth when applied to uninfected rice plants. Although it was immediately evident from this study that the fungus must be exerting its effects by secreting a potent growth stimulator, it took another 13 years to isolate the biologically active component from culture filtrates in crystalline form. Finally, in 1954 the structure of the first gibberellin, 'gibberellic acid' (now known as GA_3), was unequivocally elucidated.

During the course of this work, a variety of GA bioassays was developed to assay fungal extracts for biological activity. Once the fungal GA had been isolated, attention turned to the examination of extracts from plant tissues, particularly seeds, for gibberellin-like activity. Subsequently, the first GA from higher plant tissues was isolated from immature bean seeds by MacMillan and coworkers in 1960, and termed GA_1. Since then, over 70 naturally-occurring GAs (GA_1 to GA_{70}) have been chemically characterized, of which 24 have been identified in *Gibberella fujikuroi*, and over 60 found in different higher-plant species, some of which have also been detected in the fungus. All these compounds have the same basic carbon skeleton; however, they can be subdivided into the C_{20}-GAs which possess the complete diterpenoid complement of 20 carbon atoms, and the C_{19}-GAs in which the 20th carbon atom has been lost during their biosynthesis (Figure 2.3).

There is a general consensus that the starting point in the GA biosynthetic pathway is 3R-mevalonic acid (Figure 2.4). A sequence of reactions then takes place which involves the cyclization of geranylgeranylpyrophosphate (GGPP) to copalylpyrophosphate (CPP) and the formation of *ent*-kaurene, which is then oxidized and rearranged to give GA_{12}-aldehyde (Coolbaugh, 1983). GA_{12}-aldehyde appears to be the branch-point to the various GAs in all organisms. From this compound several different pathways have been established and these differ primarily in the position and sequence of hydroxylation.

In pea seeds, two parallel pathways have been discovered by a combination of *in-vitro* and *in-vivo* methods (Sponsel, 1985). The major pathway involves the early 13-hydroxylation of GA_{12}-aldehyde to GA_{20} and GA_{29}, and this biosynthetic sequence may be common to tissues from a wide variety of other plant species. The alternative pathway leads to GA_9 and GA_{51} (Figure 2.5). The components of these two pathways have not been found to be equally distributed throughout all the tissues of pea seeds, which implies that the GAs may be transported from one tissue to another and that particular enzymes necessary for their conversion may be restricted to specific tissues.

The early hydroxylation pathway has also been extensively investig-

(a)

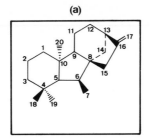

(b)

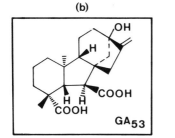

GA$_{53}$

(c)

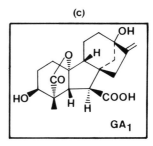

GA$_1$

Figure 2.3 Structure of (*a*) the basic gibberellin carbon skeleton; (*b*) a C$_{20}$-GA, GA$_{53}$; and (*c*) a C$_{19}$-GA, GA$_1$.

ated in maize. This study has been carried out by Phinney's group, who have used single-gene dwarf mutants to assist their work (see Figure 7.4). These mutants have genetic lesions at different points in the biosynthetic pathway of GA$_1$. The specific blockages can be elucidated by monitoring the growth of the mutants in response to the application of particular GAs and their precursors, and by radiolabelled feeding experiments where the metabolic fate of the radiolabel is followed in the maize tissues (Phinney, 1984). From their results, Phinney and co-workers have managed to place the genetic lesions of five different maize mutants at different positions in the early hydroxylation pathway. One of these mutants, *dwarf-1*, is blocked at the step immediately prior to the formation of GA$_1$. This mutant can therefore be used to evaluate the biological activity of the other GAs in the pathway. GA$_{53}$ and GA$_{20}$ can stimulate the elongation of *dwarf-1* plants, but the efficacy of these two GAs is less than 1% of GA$_1$. An important conclusion that can be drawn from this observation is that although maize shoot tissues contain at least 8 GAs, only GA$_1$ may be biologically active whilst the remainder are active through their conversion to GA$_1$. As a result of

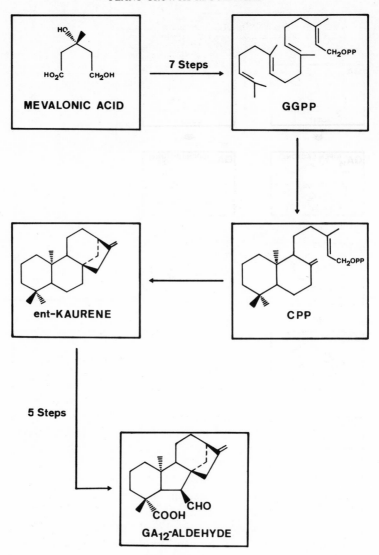

Figure 2.4 Biosynthetic pathway of the GA precursor, GA_{12}-aldehyde, from mevalonic acid (MVA). GGPP, geranylgeranyl pyrophosphate; CPP, copalyl pyrophosphate.

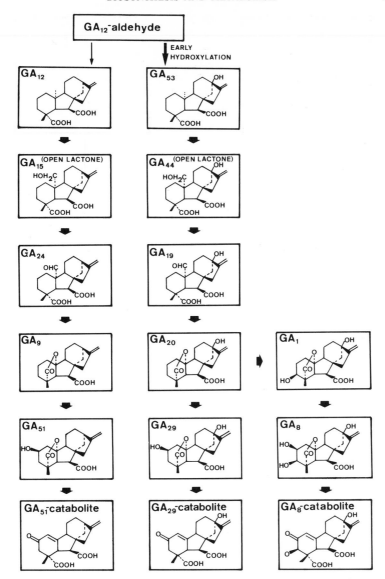

Figure 2.5 Biosynthetic and metabolic pathways of gibberellins from GA_{12}-aldehyde in *Pisum sativum* tissues. The early 13-hydroxylation pathway is the major one, and has been demonstrated to occur in a spectrum of plants.

this, it is clear that the conclusions drawn from studies which attempt to correlate the total quantity of endogenous GAs in a tissue with growth must be viewed cautiously.

Although significant progress has been made recently in the establishment of the biosynthetic pathway of GAs in some plant tissues, as yet little information is available as to how the endogenous levels of these PGRs are regulated. In the light of the conclusions from Phinney's work, this is particularly important. A potential site of regulation which has been investigated is that of the cyclization of GGPP to *ent*-kaurene, since GGPP occupies a key position both in synthesis of diterpenoids such as GAs and also of tetraterpenoids such as carotenoids. The enzyme responsible, *ent*-kaurene synthetase, has two active sites. Site A catalyses the conversion of GGPP to CPP, and site B catalyses conversion of CPP to *ent*-kaurene. One of the maize mutants, *dwarf-5*, has a lesion specifically associated with site B activity. Several reports have shown a correlation between the activity of *ent*-kaurene synthetase and shoot elongation in pea stems which suggests that the activity of the enzyme might play an important role in the regulation of GA levels. In addition, the growth-retarding properties of synthetic PGRs such as chlormequat (CCC) and AMO-1618 have been attributed to their ability to inhibit the activity of this enzyme. Since *ent*-kaurene and its immediate metabolites are apolar, it has been proposed that the enzymes responsible for their formation are associated with the endoplasmic reticulum (ER). In support of this hypothesis, it has been shown that the conversion of GA_{12}-aldehyde to GA_{12} and GA_{53} in pea tissue is catalysed by microsomal enzymes, and that the enzyme activity responsible for the oxidation of GA_{12} appears to reside on the ER. Evidence suggests that the later hydroxylating enzymes in the metabolic sequence of the GAs are soluble.

One approach which may lead to a biochemical analysis of the regulatory enzymes in the pathway is the use of transposon tagging techniques. Several GA-responding dwarf mutants of maize have already been isolated which are thought to be the result of the insertion of a transposable element into the genome resulting in the blockage of a specific step in the biosynthetic pathway (Phinney *et al.*, 1986). Since the transposon sequence is known and has been cloned, it is possible to probe restriction enzyme fragments of the maize DNA with a radiolabelled plasmid incorporating the cloned sequence to identify the specific parts of the maize genome which have been 'mutated'. This approach should lead to the isolation of the sequence of the dwarfing and normal genes, and ultimately to the biochemical analysis of the appropriate gene products responsible for the mutations.

2.2.2 Metabolism

As the number of GAs that have been identified has risen, it has become increasingly clear from studies of the biosynthetic pathways that only a small fraction of these molecules are biologically active. The remainder are either intermediates in the pathway or deactivation products. Physiological activity is believed to reside primarily with the lactonic C_{19}-GAs and is dictated by the pattern of hydroxylation of the molecules. For instance, 2β-hydroxylated C_{19}-GAs have little or no biological activity, and these compounds are commonly found as metabolites after feeding plant tissues with functional GAs (see Figure 2.5); 2β-hydroxylation of C_{19}-GAs may therefore be the natural metabolic reaction in plant tissues by which biologically active GAs are deactivated. Once formed, hydroxylated GAs may either be oxidized to unsaturated ketones such as GA_{29}-catabolite, or conjugated to form 2-O-β-D-glucosides. Glucosyl ester and ether derivatives of GA_1, GA_3 and GA_4 can also be isolated from plant tissues, but the role of these conjugates of biologically active GAs remains unknown.

As a result of studies on the distribution of GA metabolites in seeds, it has been suggested that the 2β-hydroxylating enzymes responsible for deactivating GAs may be localized in specific tissues. For instance, in maturing seeds of *Phaseolus coccineus*, GA_8 is restricted to the testa, while GA_1 is found primarily in the cotyledons. Similarly localized distributions of GA catabolites have been reported in other seeds. If this phenomenon is of general occurrence, then it would indicate that the movement of GAs from tissue to tissue has an important impact on their metabolic fate. This may turn out to be a highly significant discovery in our quest to ascertain the role of GAs in plant development.

2.3 Cytokinins

2.3.1 Background and biosynthesis

The concept that plant cell division was regulated by specific endogenous chemicals is thought to have been originally formulated by Weisner in 1892. Over 20 years later, experimental evidence to support this proposal was obtained by Haberlandt (1913), and a compound with the capacity to promote the division of tobacco pith cells was finally purified from autoclaved herring-sperm DNA by Skoog and his co-workers in 1956. This synthetic CK was identified as 6-(furfurylamino)purine, and given the name kinetin. After this success, there followed a nine-year wait before the first naturally occurring CK, 6-(4-hydroxy-3-methylbut-*trans*-2-enylamino)purine, or zeatin, was isolated from *Zea mays* kernels by Letham. Since then, other naturally occurring CKs have been isolated from a range of plant species and tissues. The majority of these compounds have proved to be derivatives of purines and they have been chemically classified as N^6 substituted adenines (see Table 2.1).

A potential source of CKs in plant cells is tRNA, since certain species of these nucleic acids contain CKs adjacent to the 3' end of the anticodon (see Figure 9.6). Hydrolysis of these tRNAs could therefore contribute free CKs to cellular pools. The significance of such a biosynthetic route is difficult to assess and has been the subject of much debate. Different tRNA species appear to turn over at different rates in plant tissues, although the average turnover rate is low and hence would not contribute significantly to free CK levels. Indeed, it has been estimated that levels of CKs in plant tissues are many times higher than the CK content of tRNA. Furthermore, the tRNA of CK-requiring tobacco callus cultures contains CKs, but this is clearly insufficient to satisfy the needs of the tissue for this PGR. On balance, therefore, the current evidence argues against a significant contribution to free CKs from tRNA.

In contrast, it is widely accepted that a substantial route for CK biosynthesis originates from adenine (Ade) metabolism. However, this is a technically difficult avenue to study because it constitutes only a minor secondary pathway of Ade metabolism. Thus, when radiolabelled Ade is supplied to plants, the vast majority of the label passes into other purines and their derivatives, and only a tiny fraction into the extremely low levels of endogenous CKs. A further approach by which to probe CK biosynthesis would be to introduce radiolabel into the CK side-chain. Mevalonic acid has been used in this way since it is a precursor of the isoprenoid side-chain of some CKs; however, the results of such experiments have not proved particularly revealing. This may be a reflection of the poor uptake of radiolabel.

The available evidence suggests that CK biosynthesis from Ade occurs at the nucleotide level and involves phosphorylated intermediates (Figure 2.6) (see McGaw *et al.*, 1984). In an attempt to characterize the pathway, investigators have sought to obtain CK biosynthesis in

Table 2.1 Structures, names and abbreviations of the most common endogenous and synthetic cytokinins. (Modified from Horgan, 1984).

Substituent groups			Common name	Abbreviation
R_1	R_2	R_3		
(2-isopentenyl, CH_3/H_2C/CH_3)	H	H	N^6 (^{2}-isopentenyl) adenine	i^6Ade
	H	ribofuranosyl	N^6 (^{2}-isopentenyl) adenosine	i^6A
(CH_2OH/CH_3/H_2C)	H	H	zeatin	io^6Ade
	H	ribofuranosyl	zeatin riboside	io^6A
(CH_2OH/CH/H_2C)	H	H	dihydrozeatin	H_2io^6Ade
(H_2C-benzyl)	H	H	N^6-(benzyl) adenine	bzl^6Ade
(H_2C-furfuryl, O)	H	H	kinetin	

cell-free systems. This was first achieved with extracts from the slime mould *Dictyostelium discoideum*, and with CK-autonomous tobacco callus cultures. In both systems, AMP accepts the isopentenyl group of ^{2}-isopentenyl pyrophosphate, and reaction (1) is believed to be catalysed by an AMP-^{2} isopentenyl transferase enzyme which has been partially purified:

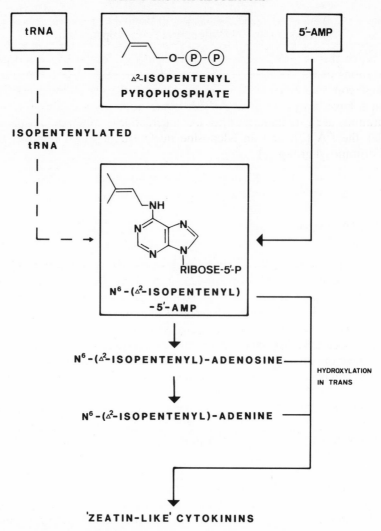

Figure 2.6 The proposed biosynthetic pathways of cytokinins. In both pathways, $\triangle^2$-isopentenylpyrophosphate plays a strategic role. This compound has been hypothesized to either condense with free 5'-AMP and then undergo further biochemical reactions, or condense with adenine residues from tRNA molecules followed by hydrolysis to active cytokinins. The current evidence suggests that the pathway of CK biosynthesis from tRNA is a minor one. (Scheme redrawn from Weiler and Schroder, 1987).

$$AMP + \Delta^2\text{-isopentenyl pyrophosphate} \rightarrow$$
$$N^6\text{-}(\Delta^2\text{-isopentenyl})\text{adenosine-5'-monophosphate} \qquad (1)$$

When this *in-vitro* reaction is performed with less well purified preparations of the transferase, the two CKs i^6A and i^6Ade are formed. These compounds are believed to be the precursors of all the other CKs which have been documented. This observation suggests that the preparations are contaminated with a 5'-nucleotidase which dephosphorylates the i^6A (2), and an adenosine nucleosidase which removes the ribofuranosyl group (3):

$$N^6 - (\Delta^2\text{-isopentenyl})\text{adenosine-5'-monophosphate} \xrightarrow{5'\text{-nucleotidase}}$$
$$N^6 - (\Delta^2\text{-isopentenyl})\text{adenosine } (i^6A) \qquad (2)$$

$$N^6 - (\Delta^2\text{-isopentenyl})\text{adenosine } (i^6A) \xrightarrow{\text{adenosine nucleosidase}}$$
$$N^6 - (\Delta^2\text{-isopentenyl})\text{adenine } (i^6Ade) \qquad (3)$$

The most abundant CKs in plant tissues are hydroxylated, which implies that, if the above reactions are indeed part of the CK biosynthetic pathway, it should be possible to achieve this type of conversion *in vitro*. In support of this hypothesis, microsomal preparations of tobacco callus cultures have been shown to contain mixed-function oxidases capable of stereospecifically hydroxylating i^6Ade to *trans*-zeatin. These enzymes also have the capacity to hydroxylate i^6A to *trans*-zeatin riboside, which in turn can be converted to *trans*-zeatin by the action of adenine nucleosidase.

Tumours induced by infection of plant cells with *Agrobacterium tumefaciens* (crown gall) contain high concentrations of CKs, and therefore are convenient tissues on which to study the biosynthesis of this PGR (see also section 2.1). The results of experiments performed on *Vinca rosea* and *Datura* crown gall tissue support the biosynthetic scheme described above, and suggest that the *trans*-hydroxylation of the isopentenyl side chain occurs so rapidly *in vivo* that only minute quantities of the intermediates can be found. Furthermore, one of the genes which regulates the CK levels in the tumorous cells has been shown to encode an isopentenyl transferase enzyme. This gene has been cloned and expressed in *E. coli* as a 27kD protein. If this protein exhibits sufficient homology to the plant enzyme, it may be possible to use antibodies raised against it to probe the sites of CK biosynthesis in plant tissues. Alternatively, *in-situ* hybridization with the cDNA probe itself could be used to identify those cells and tissues where the gene is expressed.

2.3.2 Metabolism

Plant tissues have the capacity to rapidly metabolize both synthetic and naturally occurring CKs (see McGaw et al., 1984). Some of the metabolites formed, for instance those generated by hydroxylation of the side-chain, or conjugation to sugars or amino acids, retain some degree of CK activity. Others are the result of cleavage of the side-chain and hence have no biological activity. An enzyme which may play a role in the generation of the latter group of metabolites is CK-oxidase. This enzyme can cleave the isopentenyl group from such substrates as i^6A, i^6Ade, zeatin and zeatin riboside. Although this reaction could contribute significantly to the regulation of endogenous CK concentrations, it has been shown that crown gall-infected Vinca rosea tissue which has a high CK level also exhibits highly elevated CK-oxidase activity. This paradoxical observation emphasizes the point that the fate of exogenously applied CKs need not mimic the situation in vivo.

Conjugation of PGRs with sugars or amino acids is a common metabolic phenomenon. The structure of CKs allows conjugation to take place at three positions on the purine ring. The most widespread conjugates are the 9-ribosides and their 5'-monophosphates. Glucosides are also common; however, the glucose moiety is not restricted to the 9-position and conjugation can also take place at the 3- and 7-sites. Side-chain conjugated O-glucosides have also been found in Lupinus sp. and Vinca rosea crown gall tissue. In a variety of species, amino acid conjugates can be isolated after feeding CKs such as zeatin to plant tissues. Most commonly the residue is alanine, and this is located at the 9-position on the purine ring.

The brief picture of CK metabolism that has been painted demonstrates the complexity of the process. A further complication is that the pattern of metabolites that can be identified in different tissues and species varies considerably. What therefore can be concluded about the role of CK metabolites in the regulation of endogenous CK levels in plant cells? One suggestion that has been put forward is that CK bases such as zeatin require metabolic conversion before they can act as regulators of cell division. However, in the absence of convincing experimental support this proposal has now fallen from favour. Other possibilities are that some metabolites act as storage forms of CKs whilst others are inactivation products. It is clear that if we are to ascribe a role to individual metabolites, a great deal of further work is necessary. For instance, the enzymes involved in CK metabolism must be identified, and the cellular and subcellular sites at which they act must be localized.

The task of further probing CK biosynthesis or metabolism would be greatly facilitated with the aid of mutants. As yet, the only CK mutant that has been characterized is a mutant of the moss *Physcomitrella patens* which has elevated endogenous CK levels. Whether the genetic lesion associated with this mutant lies at the level of biosynthesis or metabolism is presently unknown. PGR-deficient mutants have proved of considerable assistance in the elucidation of the biosynthetic pathway of GAs and ABA. Clearly, CK-deficient mutants would also be of great value; however, if CKs play a significant role in the regulation of plant cell division, such mutations might need to be very 'leaky' in order to ensure their viability, and hence they might be difficult to recognize.

2.4 Abscisic acid

2.4.1 Background and biosynthesis

The systematic search for an endogenous plant growth inhibitor was initiated by the work of Bennet-Clark and his colleagues in the early 1950s. They discovered that the acid fraction of an alcoholic extract from a variety of plant tissues displayed growth inhibitory properties. This fraction was termed 'inhibitor β', and prompted a number of research groups to focus their attention on its identification. Little progress was made until the early 1960s, when a series of publications appeared reporting the isolation of compounds which had the ability to inhibit growth of coleoptile segments. These compounds were given the trivial names 'dormin', and 'abscisin II', since they had been obtained from dormant sycamore buds (Wareing *et al.*, 1965), and dehiscing cotton bolls (Addicott *et al.*, 1965), respectively. The molecular structure of these two compounds was found to be identical, and this 'putative' growth inhibitor was christened abscisic acid (ABA). Whether ABA was the major biologically active constituent of Bennet-Clark's 'inhibitor β' remains a matter of conjecture.

ABA is a sesquiterpenoid composed of three isoprene residues. In addition, the carbon skeleton of the molecule bears a close similarity to the terminal rings of carotenoids such as violaxanthin. It can exist in either the *trans* or *cis* configuration, and the latter, being optically active, has a (+) and (−) enantiomer. Both the *cis* and the *trans* isomers of ABA may be extracted from plant tissues, although only the former isomer exhibits biological activity. Light isomerizes ABA to a mixture of

B

the *cis* and *trans* structures, and this photolytic conversion is thought to account for the presence of the inactive isomer in plant extracts. In view of the structure of ABA, two potential pathways for its biosynthesis have been proposed. The first is a 'direct' pathway in which a C_{15} precursor, possibly farnesyl pyrophosphate, is directly converted to ABA. The alternative or 'indirect' pathway assumes that ABA originates from the enzymic cleavage of a C_{40} xanthophyll to yield a C_{15} intermediate which is metabolized to give the PGR.

Attempts to distinguish between the two pathways have, until recently, centred on the feeding of radiolabelled 'putative' intermediates to plant tissues and the assessment of their efficiency of conversion to ABA. The results of such studies demonstrate that compounds such as the terpenoid precursor mevalonic acid can be incorporated into ABA; however, the incorporation is very small and can be as low as 0.001%. Furthermore, in either pathway, mevalonic acid would be the ultimate precursor.

Recently, as a result of two independent lines of investigation, the balance of evidence has strongly tipped in favour of the indirect pathway. The first has utilized the ability of water stress to stimulate ABA biosynthesis (see Chapter 4). Work by Creelman and Zeevart (1984) has shown that if plants are stressed in the presence of $^{18}O_2$, the ABA that is synthesized incorporates the isotope into only a single oxygen atom within the carboxylic group. This observation is most compatible with the indirect pathway, since oxygen would only be a prerequisite for the cleavage of the C_{40} carbon skeleton. If the direct pathway was operative, one would predicate that the ring structure would also become labelled in such an experiment.

The other approach has utilized PGR-deficient mutants. ABA-deficient mutants can be readily characterized by their 'wilty' phenotypes. Three such mutants have been identified in tomato and these have been termed *flacca*, *notabilis*, and *sitiens* (see Figure 8.1). Working on the hypothesis that the mutant gene loci of these 'wilty' plants are involved in coding for enzymes in the biosynthetic pathway of ABA, Taylor and co-workers have been searching for intermediates which might accumulate behind the genetic lesions. The work of this group has led to the isolation of two compounds of interest. The first is a C_{10} compound, 2,7-dimethyl-2,4-octadienedioic acid (ODA). This compound has been proposed to be a byproduct of the indirect pathway of ABA biosynthesis. The logic behind this proposal is that cleavage of one molecule of a C_{40} precursor would generate two molecules of a C_{15} ABA intermediate and a C_{10} by-product. In support of this hypothesis is the demonstration that *notabilis* has an unusually low concentration of ODA, which suggests that the mutation acts prior to the cleavage step. In contrast, *flacca* and *sitiens* have abnormally high concentrations of ODA, which implies that the genetic lesions affect one of the steps in the biosynthetic pathway subsequent to cleavage (for a more detailed discussion see Taylor, 1987). Recently, a second compound has been found to accumulate in the *flacca* and *sitiens* mutants following water stress. This has been identified as *trans*-ABA alcohol (Linforth *et al.*, 1987). The *trans*-ABA alcohol level in these plants has been calculated to be approximately equal to the deficiency in ABA. Although the simplest explanation to account for this observation is that *trans*-ABA alcohol is an intermediate in the biosynthetic pathway, feeding experiments have revealed that plants are unable to convert this compound to ABA. One plausible explanation for this apparent paradox is that the immediate precursor of this PGR can be converted *in situ* to either ABA or *trans*-ABA alcohol, and that although the former reaction predominates in wild-type plants, in *flacca* and *sitiens* this step is blocked and therefore *trans*-ABA alcohol accumulates (Figure 2.7).

Further evidence in support of the indirect pathway has come from experiments where seedlings have been exposed to inhibitors of carotenogenesis. If ABA does originate from the cleavage of a xanthophyll, it would be predicted that such inhibitors would reduce the capacity of plant tissues to accumulate the PGR. The herbicides fluridone and norflurazon block carotenoid biosynthesis by inhibiting the desaturation of phytoene to phytofluene. Seedlings grown in the presence of these herbicides have been shown to have greatly reduced ABA levels.

2.4.2 Metabolism

ABA can be readily metabolized by plant tissues, resulting in the formation of several conjugates and metabolites. It is evident from feeding experiments that the (+) and (−) enantiomers are not metabolized in the same way. The products of metabolism of the natural (+) enantiomer have been placed in a proposed metabolic sequence by Loveys and Milborrow (Figure 2.8). In this scheme it is suggested that ABA is initially oxidized to form the unstable compound 6′-hydroxymethyl-ABA which 'rearranges' to give phaseic acid (PA). High rates of conversion of ABA to PA have been reported in a variety of tissues; however, this step has only been observed *in vitro* utilizing a cell-free system from

CAROTENOID PRECURSORS

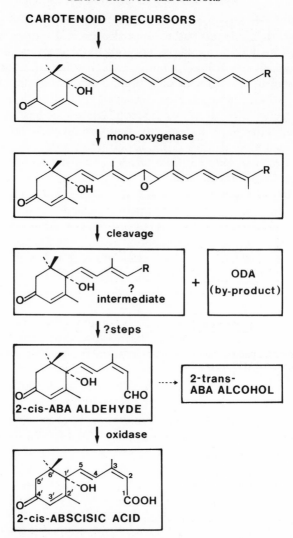

Figure 2.7 Hypothetical scheme for the biosynthesis of ABA. Current opinion favours the view that ABA is synthesized by an indirect route from carotenoid precursors. The precise intermediates in the pathway are unknown; however, there is good evidence to believe that carotenoids are cleaved to generate two C_{15} molecules and the C_{10} fragment 2,7-dimethyl-2,4-octadienedioic acid (ODA). Recent work suggests that 2-*cis*-ABA-aldehyde is the immediate precursor of ABA, and that if this reaction is blocked, 2-*trans*-ABA alcohol accumulates. (Scheme modified from Linforth *et al.*, 1987, after consultation).

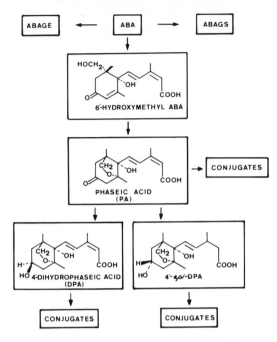

Figure 2.8 Proposed pathways of ABA metabolism. ABAGE, (+)-abscisyl-β-D-glucopyranoside; ABAGS, 1′-O-abscisic acid-β-D-glucopyranoside. (After Loveys and Milborrow, 1984.)

Echinocystis lobata liquid endosperm. The majority of plant tissues do not accumulate PA but subsequently convert it to 4′-dihydrophaseic acid (DPA) or the epimer *epi*-DPA. In general, ABA metabolites exhibit little biological activity. An exception to this is PA which successfully mimics the effect of ABA in inhibiting GA-induced secretion of α-amylase in barley aleurone layers (see Chapter 6), and displays a limited ability to promote leaf abscission in cotton seedlings (see Chapter 8).

If plant extracts are hydrolysed by alkali, the free ABA content of the extracts is increased. The source of this ABA is ABA-conjugates. At least two conjugates have been identified in plant tissues. The most prevalent compound is the glucose ester of ABA (ABAGE); however, a second conjugate, the 1′-O-glucoside (ABAGS), has also recently been characterized. There is no evidence that these conjugates act as a source of free ABA, since wilted plants accumulate ABA in the absence of a change in levels of ABA conjugates. Furthermore, *Beta vulgaris* plants that have been pretreated with ^{14}C-ABA to radiolabel ABA conjugates do not accumulate an enhanced amount of free ^{14}C-ABA on water stressing. The compounds phaseic acid and dihydrophaseic acid can also be conjugated, and both the glucose esters and the glucosides of these ABA metabolites have been identified in plant tissues.

It is clear that good progress has been made towards the identification of intermediates in the biosynthetic pathway of ABA. Once again the use of plant mutants in the study of PGR biosynthesis is proving to be very valuable. The site of the genetic lesions responsible for these mutant phenotypes remains to be determined; however, they may reside in enzymes which are critical regulators of the pathway. As new mutants are identified, it should be possible to use them to characterize additional steps which occupy strategic positions in the biochemical sequence. It is evident that metabolism can also play a fundamental role in the regulation of ABA concentrations in plant cells. For instance, water stress not only stimulates ABA biosynthesis, but also metabolism, and the return of ABA to its basal level after the alleviation of stress is a function of a reduction in biosynthesis and the maintenance of elevated metabolism (see Chapter 4). The current approach to identifying the products of ABA metabolism and the enzymes responsible for their conversion, has centred on the use of feeding experiments, and the results are therefore subject to all the provisos highlighted previously. An alternative method would be to generate mutants and screen for phenotypes which might exhibit an elevated or depressed capacity to metabolize ABA. Such metabolic mutants, like biosynthetic mutants, could have a dramatic impact on our understanding of the regulation of PGR levels in plant cells.

2.5 Ethylene

2.5.1 Background and biosynthesis

The proposal that volatiles could have a profound effect on plant development originates from the observation made by Girardin (1864) that leaking illuminating gas could promote leaf abscission. Twenty years later, Molisch noted that there was a correlation between the abnormal gravitropic responses of seedlings and the presence of traces of illuminating gas in the atmosphere. Similar observations were made by Neljubow (1901) who used the 'triple response' of etiolated pea epicotyls (diageotropism, inhibition of elongation, stimulation of lateral expansion), to assay the various components of coal gas for biological activity. The constituent with the greatest activity was found to be ethylene, and the gas was positively identified as a natural plant product by Gane in 1934.

It has been confirmed that ethylene has a significant impact on developmental processes throughout the life cycle of a plant, including germination, growth, ripening, senescence and abscission. Moreover, production of the gas is probably a ubiquitous feature of plant cells, although the ethylene biosynthetic capacity of a cell may vary during its ontogeny. Many compounds have been proposed as potential precursors of ethylene; however, only methionine can be readily converted to ethylene by plant tissues *in vivo*. It has now been established that ethylene originates from the C3 and C4 positions of the methionine molecule, and that the CH_3S group is retained by the plant tissue and recycled. This recycling mechanism has strategic significance for plant cells: they can sustain a steady rate of ethylene production even when their endogenous levels of methionine are low. The mechanism is mediated through the production of the intermediate S-adenosylmethionine (SAM), which fragments to form methylthioadenosine (MTA). MTA is rapidly hydrolysed to methylthioriboside (MTR) and thence converted back to methionine.

In an anaerobic environment, the production of ethylene by plant tissues is inhibited. On aeration, the tissue produces a large 'pulse' of ethylene. This phenomenon was investigated by Adams and Yang (1979), who compared the metabolism of ^{14}C-methionine by apple slices under aerobic and anaerobic conditions. In air, the apple tissues readily converted methionine to ethylene; however, in nitrogen, the tissues accumulated a radiolabelled compound which when fed to aerated tissue could be metabolized to ethylene. Adams and Yang reasoned that this unknown compound might be an ethylene precursor which required oxygen for further conversion. They therefore purified the compound, and it was identified as 1-aminocyclopropane-1-carboxylic acid (ACC). ACC has now been detected in a wide variety of plant tissues, and has become universally accepted as the immediate precursor of ethylene. Apart from conversion to ethylene, it has also been demonstrated that ACC can be metabolized to the non-volatile compound N-malonyl-ACC (MACC). The role of this pathway is unknown, however, since MACC exhibits little turnover or reconversion to ACC, it represents a potential mechanism for regulating ethylene production.

Whilst the nature of the intermediates in the biosynthetic pathway

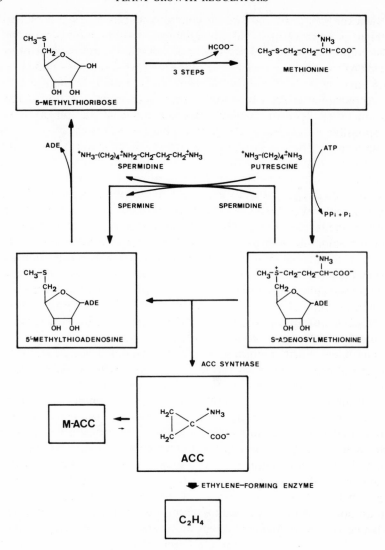

Figure 2.9 Biosynthetic pathway of ethylene showing its relation to polyamine synthesis in plant tissues. ACC, 1-aminocyclopropane-1-carboxylic acid; M-ACC, N-malonyl-ACC; ADE, adenine.

of ethylene is well established (Figure 2.9), little is known about the enzymes associated with the pathway. The enzyme which catalyses the conversion of SAM to ACC, ACC synthase, has recently been puri-fied from tomato pericarp tissues by immunoaffinity chromatography

(Bleecker *et al.*, 1986). The enzyme is reported to have a molecular weight of approximately 50kD. Activity of the enzyme is supressed by the compounds aminoethoxyvinylglycine (AVG), and aminooxyacetic acid (AOA), both of which are inhibitors of pyridoxal phosphate requiring enzymes.

Application of ACC to many plant tissues results in a rapid stimulation of ethylene production. This suggests that the enzyme system responsible for catalysing the conversion of ACC to ethylene, the ethylene-forming enzyme system (EFE), is largely constitutive and that the formation of ACC is the major rate-limiting step in the biosynthetic pathway. The activity of the EFE can increase during developmental events such as ripening, and in response to ethylene itself; in addition, it may be influenced by environmental conditions such as stress. There is convincing evidence that the ethylene-forming system is dependent upon membrane integrity, and it has been proposed that the conversion of ACC to ethylene may be coupled to a transmembrane flow of protons from the outside to the inside of the plasmamembrane. Progress on the intracellular localization of ethylene biosynthesis has been slow. It has been reported that over 80% of the ACC found in protoplasts from pea leaves is found in the vacuole. Furthermore, the characteristics of the ethylene-forming system of isolated vacuoles resembles that of the intact tissue. Further studies are necessary to determine whether this is a general property of plant vacuoles, and to localize the site of synthesis of ACC.

During the oxidation of ACC to ethylene, the amino nitrogen is released and becomes incorporated into asparagine via the intermediate β-cyanoalanine. Since one of the initial products of the EFE reaction is the highly toxic CN^-, detoxification may be mediated by the conversion of CN^- to β-cyanoalinine via the action of β-cyanoalanine synthase. Support for this hypothesis comes from the recent demonstration that the ethylene climacteric which is associated with the senescence of carnation flowers is accompanied by an increase in the activity of β-cyanoalanine synthase (Manning, 1986).

2.5.2 *Metabolism*

Studies on the capacity of plant tissues to metabolize ethylene were hampered for a number of years by the lack of radiochemically pure ^{14}C-ethylene. In 1975 this problem was overcome by Beyer, who incubated etiolated pea seedlings in ^{14}C-ethylene that had been vigorously purified. His work showed that, while rates of ethylene metabolism by pea tissues were low, significant radioactivity was incorporated into CO_2 and products that were retained within the tissue. These products were later identified as ethylene glycol and the glucose conjugate β-2-hydroxyethyl-D-glucoside. Ethylene metabolism has now been demonstrated in a range of tissues and plant species (see Smith and Hall, 1984). In general, rates are comparable to those found in pea seedlings; however, cotyledons of *Vicia faba*, and alfalfa seedlings, exhibit rates that are of an order of magnitude higher. In addition, in *Vicia faba*, only one major product is formed, and this has been identified as ethylene oxide. Since this compound can be readily converted to ethylene glycol either enzymically or non-enzymically, it is likely that the initial metabolite incorporated into pea tissues is also ethylene oxide. The enzyme responsible for metabolizing ethylene to ethylene oxide has been characterized in an extract from *Vicia faba* cotyledons. The system has a high affinity for ethylene (K_m of 10^{-9} M), and requires both oxygen and the cofactor NADPH for activity. These properties, coupled with the observation by Beyer that *Vicia faba* cotyledons produce radiolabelled ethylene oxide in the presence of $^{18}O_2$, suggest that the enzyme is a monooxygenase.

It has been frequently postulated that metabolism is one method by which tissues can remove or inactivate excess PGRs. In the case of ethylene, there are a number of reasons why this hypothesis may need to be reconsidered. Firstly, as it is a gas, ethylene readily diffuses out of most tissues. Secondly, apart from *Vicia faba* and alfalfa seedlings, the amount of metabolism in plant tissues would appear to be too small to act as an effective method of ethylene control. Thirdly, compounds such as carbon disulphide inhibit ethylene metabolism, but have no discernible effect on the amount of the gas produced by the plant system.

A clue to the possible role of ethylene metabolism in plants has come from studies of the sensitivity of different tissues to ethylene and their ability to metabolize the PGR. For instance, the capacity of *Ipomea tricolor* flower buds to respond to ethylene increases in concert with the ability to metabolize the gas. Comparable changes have been observed in leaf abscission zones from cotton plants. In addition, in etiolated pea seedlings, the potency of silver ions (which are hypothesized to interfere with ethylene binding to its cellular receptor — see Chapter 9) to inhibit ethylene action and metabolism exhibit a remarkably similar profile when plotted against increasing ethylene concentration. On the basis of these observations, Beyer (1985) has proposed that ethylene metabolism may be an important prerequisite for ethylene action.

The mechanism by which metabolism and action might be linked is unknown. One hypothesis is that ethylene oxide can modulate the interaction of ethylene with a cellular receptor. This idea has stemmed from the demonstration that ethylene oxide has a synergistic effect with ethylene on the induction of the 'triple response' in peas, and leaf abscission in cotton. Ethylene oxide does not stimulate these responses in the absence of ethylene, nor does it elevate production of the PGR. This hypothesis does not invoke a strategic role for the CO_2-forming component of ethylene metabolism, although it might act to increase the activity of the ethylene monooxygenase (Sanders *et al.*, 1986).

In recent years considerable progress has been made in the identification of the intermediates in the biosynthetic and metabolic pathways of ethylene. However, as yet our knowledge of the enzymes involved in these biochemical pathways is meagre. Clearly, this information is important in order to understand how ethylene levels are regulated in plant tissues. However, if the hypothesis that ethylene metabolism and action are linked is correct, knowledge of the enzymic regulation of metabolism assumes an even greater significance. The elucidation of these biochemical reactions may be a crucial step towards the goal of understanding how ethylene regulates plant development.

2.6 Polyamines

The chemical structure of the polyamines spermidine and spermine was established in the 1920s, but the significance of these compounds in plant growth and development has only been recognized over the last 10 years. During this time it has become established that polyamines play a significant role in embryogenesis, germination, cell growth, senescence, and the response of plants to stress. In addition, there is an increasing amount of evidence to suggest that some of the effects of PGRs are mediated through changes in polyamine biosynthesis or metabolism (Smith, 1985).

Polyamines can be synthesized sequentially from the diamine putrescine. In this reaction, one or two aminopropyl groups are donated from decarboxylated S-adenosylmethionine to give spermidine or spermine respectively. Putrescine may be formed directly by the decarboxylation of ornithine, or alternatively through a series of intermediates following arginine decarboxylation. The other precursor originates from the compound S-adenosylmethionine (SAM), which plays an integral role in the regulation of ethylene biosynthesis (see section 2.5). Since the biosynthetic pathways of ethylene and the polyamines compete for a common precursor, a shift in the metabolism of one could have a direct effect on the endogenous levels of the other (see Miyazaki and Yang, 1987). Certainly it has been reported that the chemical inhibition of SAM decarboxylase or arginine decarboxylase stimulates the ethylene production by carnation flowers, while the inhibition of ethylene biosynthesis in the same tissue is accompanied by an increase in spermine biosynthesis.

Polyamine levels in plant tissues may also be regulated by metabolism. The enzyme diamine oxidase has now been identified in a wide spectrum of plant tissues, and this has the capacity to oxidize a range of substrates including putrescine or spermidine. The oxidation product of putrescine, pyrroline, may be further oxidized to 4-aminobutyric acid. Another enzyme, polyamine oxidase, specifically oxidizes spermidine and spermine; however, this enzyme has so far only been detected in the Gramineae.

The ubiquitous nature of spermidine and spermine and their growth regulatory properties suggest that these compounds may have an important role in plant development. As yet our knowledge of their metabolic pathways is fragmentary. However, from the information that is available already, it is clear that the regulation of these processes can have a significant impact on the endogenous levels of PGRs such as ethylene. It remains to be seen whether polyamines play the role of secondary messengers for the major groups of PGRs, or can themselves be classified as a group of molecules with plant growth regulator properties.

2.7 More than making and breaking

In this chapter we have briefly described the history of the discovery of the major PGRs and outlined their biosynthesis and metabolism. Clearly, encouraging progress has been made in attempts to elucidate the biosynthetic pathways of the GAs and ethylene, and researchers may be on the brink of unravelling the intermediates of ABA biosynthesis. However, the biosynthetic routes for both IAA and CKs remain open for debate. In addition, our knowledge of the metabolism of PGRs is at best fragmentary. Nevertheless, we should not be too despondent over our lack of understanding of PGR biosynthesis and metabolism. As new techniques for extraction and assay become available, and as an increasing number of developmental mutants are characterized, future progress in the identification and significance of compounds resulting from the 'making and breaking' of PGRs should be rapid.

In striving to understand biosynthetic and metabolic pathways, we are seeking to identify also the crucial regulatory steps inherent in them and those key enzymes that mediate rate-limiting steps or crucial branch-points. Already, encouraging progress has been made towards the identification and purification of some enzymes which play a strategic role, and these studies are paving the way for exciting future research into *more* biochemical and molecular aspects of PGR biosynthesis and metabolism. Armed with this knowledge and the skills of recombinant DNA technology, it can be foreseen that in the near future the activities of crucial PGR biosynthetic or metabolic enzymes will be manipulated in transgenic plants for the benefit of the agricultural or horticultural industries.

CHAPTER THREE

EXTRACTION, IDENTIFICATION AND QUANTIFICATION — THE STATE OF THE ART

Attempts to elucidate the role of auxins, GAs, CKs, ABA and ethylene in differentiation and development have relied heavily upon the determination of the temporal and spatial changes in the endogenous level of these compounds in plant tissues. This is a technically challenging task, because the concentration of these compounds is often 1000 times smaller than that of other metabolites within a plant cell, many of which also have the capacity to produce a signal in the detectors employed for PGR analysis. Therefore, meaningful data can currently only be obtained by the extraction and partial purification of PGRs prior to their identification and quantification. This chapter examines the techniques utilized in PGR analysis, and considers the significance of the estimations obtained in understanding plant growth and development.

3.1 Extraction

During the extraction of PGRs, plant material is homogenized to ensure maximum cellular disruption. This 'crude' approach maximizes PGR recovery, but minimizes the information content that can be obtained about inter- and intracellular compartmentation. Fresh tissue can be homogenized, but more commonly it is frozen in liquid nitrogen and ground to a powder. Freezing both facilitates homogenization and reduces the risk of enzymic breakdown or modification. Traditionally, aqueous (80%) methanol at 0–4°C has been used to extract PGRs, but the basis for this has little foundation in the literature since few studies have been undertaken to examine the efficacies of other solvents. In particular, the use of aqueous methanol to extract CKs has been questioned because phosphatases are active in this solvent. The use of Bieleski's mixture (methanol, chloroform and formic acid) overcomes this problem, thus minimizing degradation of CK ribotides to ribosides. However, care must be taken when adopting complex solvent mixtures because they can solubilize compounds which may interfere with analysis methods such as immunoassay.

The cellular complement of PGRs comprises both free molecules and forms conjugated to sugar or amino acid moieties. Clearly the relative levels of free and conjugated forms of a PGR *in vivo* may be highly significant. Therefore, extraction methods should seek to restrict both the hydrolysis of conjugates and the conjugation of free PGRs. To date, considerations of these conversions have been rare.

Because it is a gas, ethylene poses different extraction problems from the other PGRs. It readily diffuses out of plant tissues and can be collected by incubating material in a gas-tight container. Samples can then be withdrawn for analysis. The beauty of this method is that it is non-destructive, and by sequential sampling, determinations of the emanation of ethylene can be made during the course of a developmental event. On the other hand, the forced accumulation of ethylene and the restriction of gaseous exchange may influence the natural progress of the developmental phenomenon under study. These problems can be solved by the use of a flow system of humidified air passing over the plant material, and trapping out evolved ethylene with absorbants such as mercuric perchlorate for subsequent analysis (see Ward *et al.*, 1978). Neither of these techniques determines the level of ethylene dissolved in the internal aqueous phase, and this can only be obtained by a vacuum extraction method. Although the technicalities behind extraction of ethylene are simpler than those of the other PGRs, production by plant tissues can be readily supplemented by emanations from materials such as plastics. As a general precaution, prior to ethylene analysis all enclosing vessels should be scrutinized for the liberation of contaminating volatiles.

3.2 Purification

Crude PGR extracts are purified by solvent partitioning and chromatography. The solvents most commonly employed include chloroform, diethyl ether and ethyl acetate. Purification procedures are designed to reduce contamination by interfering molecules, but by their very nature reduce PGR recovery. To correct for this loss, the extract is 'spiked' prior to purification with a known quantity of an internal standard. Two types of internal standard have been employed. The first is the purified PGR labelled with ^3H or ^{14}C to allow independent quantification at the end of the extraction procedure. The reliability of this approach is dictated by the degree of degradation of the standard which takes place during the clean-up process. Ideally, this should be the same as the PGR being purified. The alternative method relies on access to pure PGRs which have been labelled with heavy isotopes such as ^2H or ^{13}C. The relative amounts of standard and PGR in the final extract are determined by mass spectrometry. The advantage of this procedure is that the method of quantification takes losses during purification into account.

Solvent partitioning permits bulk extracts to be reduced to more manageable volumes while at the same time removing contaminating material. A variety of techniques have been developed for preliminary clean-ups of plant extracts, and the reader is referred to Yokota *et al.* (1981) for a detailed exposition of the relative merits of these.

Following solvent partitioning, PGRs can be further purified by chromatographic procedures. Paper- and thin-layer chromatography have, for many years been employed for this step, and these techniques continue to find use due to their low cost and ease of implementation. Unfortunately, however, these methods have limited resolution and recoveries are highly variable. Open columns for PGR purification have proved particularly effective, and partition, ion exchange, adsorption and gel filtration columns have all been employed with degrees of success dependent upon the properties of the compound to be purified. Open columns have now been largely superseded by disposable cartridges which are packed with a range of materials tailored to suit the scale of clean-up required.

The most powerful methods of purification currently available utilize gas chromatography (GC) or high performance liquid chromatography (HPLC) systems in preparative modes. The eluent from a preparative GC is split. Half is coupled to a thermal conductivity or flame ionization detector, and the other half passes through a cooled trap, where appropriate volatiles are condensed out. More commonly, HPLC systems are employed which can separate compounds having only subtle differences in molecular structure. Reverse phase HPLC is particularly powerful in this respect, and because it can be used with aqueous mobile phases, it is especially suited for PGR purification.

3.3 Identification and quantification

3.3.1 Bioassays

Historically, bioassays have played a central role in botanical research, since they have been instrumental in the discovery and isolation of every major group of PGR. Many bioassays have been developed, but for acceptance into general use they must be reproducible, sensitive and exhibit a high degree of specificity. In practice these specifications are often difficult to achieve, especially if crude extracts are assayed for PGR activity, since these may contain compounds which interfere synergistically or antagonistically with the assay. For this reason, a wide variety of assays based on different biological responses have been developed. Parameters such as cell elongation, cell division, pigment accumulation or enzyme activity have all been employed to assay for different groups of PGRs. The most frequently adopted bioassays and their sensitivities to applied PGRs are listed in Table 3.1. The threshold of PGR detection in different bioassays varies considerably. For instance, members of the GA family can invoke widely disparate responses when screened in different bioassays. This property may be turned to advantage and used to distinguish between the presence of different GAs within a single plant extract.

It is evident that the standards currently expected in biochemical analysis cannot be rigorously achieved with the sole use of bioassays. In experienced hands, however, bioassays can yield valuable information which assists significantly in the quest for quantitative data by the

Table 3.1 Bioassays for plant growth regulators and their sensitivity (adapted from Reeve and Crozier, 1980)

Plant growth regulator	Limit of detection (approximate)	Linear range (approximate)
Auxin (IAA)		
Avena curvature	$10ng\ ml^{-1}$	$10-100ng\ ml^{-1}$
Cress root growth	$1pg\ ml^{-1}$	$10ng-1\mu g\ ml^{-1}$
Gibberellin (GA$_3$)		
Barley α-amylase half seed	$10pg\ ml^{-1}$	$10pg-100ng\ ml^{-1}$
Dwarf rice leaf sheath	$10pg\ ml^{-1}$	$10pg-100ng\ ml^{-1}$
Cytokinin (kinetin)		
Radish cotyledon	$10ng\ ml^{-1}$	$10ng-1\mu g\ ml^{-1}$
Amaranthus betacyanin	$50ng\ ml^{-1}$	$50ng-5\mu g\ ml^{-1}$
Abscisic acid		
Commelina stomatal aperture	$10pg\ ml^{-1}$	$10pg-10\mu g\ ml^{-1}$
Lemna growth	$1pg\ ml^{-1}$	$1pg-100ng\ ml^{-1}$
Ethylene		
Pea triple response	$10nl\ l^{-1}$	$10-100nl\ l^{-1}$

utilization of a more 'sophisticated' technique. Furthermore, as isogenic mutants with PGR deficient genotypes become characterized, novel bioassays may be developed, exhibiting greatly refined specificities.

Bioassays have proved to be a highly effective means of isolating new PGRs. In the adoption of this method, samples are screened for activity by direct application to plant tissues. An inherent weakness of this approach is that it takes no account of the problems associated with the uptake and metabolism of a putative PGR. Also, the technique only allows the introduction of samples into the apoplastic environment surrounding a plant cell. If the movement of a PGR was restricted by its molecular structure to the symplasm, it would be exceptionally difficult to devise a conventional bioassay for it. If this line of argument is pursued, it is evident that families of PGRs with critical regulatory functions may as yet remain undiscovered.

3.3.2 Chemical methods

Although a number of chemical methods have been developed to quantify PGRs and in particular auxins, only the spectrofluorimetric method for IAA analysis has become adopted into regular practice. This assay is based on the generation of the highly fluorescent derivative indolo-α-pyrrone which has a characteristic fluorescence spectrum. Unfortunately the technique is subject to a number of constraints, the most severe of which is the quenching attributable to the presence of impurities within some plant extracts. Attempts have been made to overcome this problem by modifying the original procedure. Separation of the chemical products by HPLC and detection by fluorimetry has also improved significantly the reliability of the assay.

3.3.3 Immunological methods

Based on the proven success of immunological techniques for animal hormone assays, a number of immunological assays have been developed to quantify PGRs (Weiler et al., 1986). These methods rely on the generation of antibodies to PGRs which have been coupled to carrier molecules such as bovine serum albumin (BSA) to enhance their antigenicity. A percentage of the antibodies raised in this manner specifically recognize the hapten (PGR) component of the conjugate. By coupling the carrier to the PGR molecules at different sites, it is possible to generate antibodies exhibiting different selectivities. This approach has been successfully used to produce antibodies capable of distinguishing between free and conjugated forms of ABA (Mertens *et al.*, 1983). The selectivity of the procedure can be further enhanced by the adoption of monoclonal antibody technology. In this manner a continuous supply of antibodies with known characteristics can be obtained. Recently, monoclonals capable of recognizing a number of specific sites within the GA molecule have been produced. These have been shown to have the capacity to discriminate between the different classes of GAs (Knox *et al.*, 1987).

Quantification using antibodies can be achieved by either radioimmunoassay (RIA) or enzyme-linked immunoassay (ELISA). These assays rely on the ability of the sample antigen to compete with a known quantity of labelled antigen for a limited number of binding sites (Figure 3.1). The choice of a radiolabelled antibody may enhance the precision of the assay, but it is generally less sensitive than ELISA and not as suited to routine analysis on a large scale. The major advantage of immunoassay over other physicochemical techniques of PGR analysis is that in theory it should offer maximal specificity with minimal interference from extraneous compounds. Extensive purification should therefore be unnecessary. In practice, assay responses are modulated by the presence of a high level of 'noise' in some plant extracts, therefore a limited purification procedure is necessary.

In addition to offering a method of quantification, antibodies to PGRs can be used in other ways. For instance, immunoaffinity columns are proving to be a powerful tool for PGR purification (Davis *et al.*, 1986). In this technique, antigenic molecules are 'encouraged' to bind to antibodies immobilized to a column support, and can therefore be separated from co-extracted compounds. Bound antigens are subsequently eluted from the column under conditions which favour dissociation of the immune complex. The recovered antigen (PGR) can then be definitively identified and quantified by a method such as combined gas chromatography–mass spectrometry (GCMS). Antibodies can also be employed in localization studies (Perrot-Rechenmann and Gadal, 1986), which is the logical progression in PGR analysis. With these probes it should be possible to elucidate not only the quantity of a particular PGR within a plant tissue, but also its precise cellular and subcellular distribution. An exciting step towards the achievement of this latter goal is the development of lasers as fluorescence inducers in a variety of fluorimetric

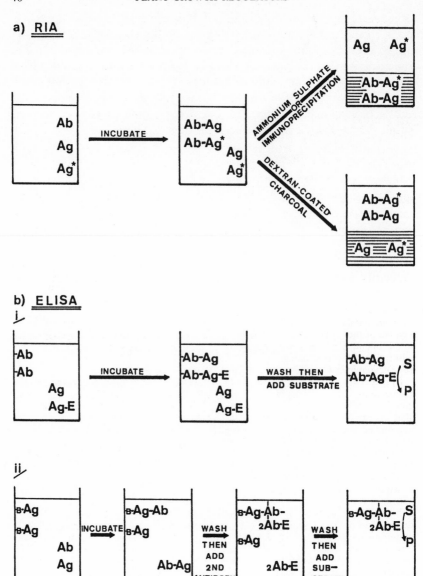

detection systems. This technology could open the way to the detection and localization of individual molecules of PGRs within a cell by fluoroimmunoassay.

3.3.4 Gas chromatography

Separation of volatile compounds can be readily achieved using gas chromatographic techniques. Furthermore, with the aid of specific detectors the compounds may be identified and quantified. At first sight this technique would appear to have limited application to PGR research, since at room temperature and pressure only ethylene is gaseous. However, by simple derivatizations all the other PGRs can be made amenable to GC analysis. In general, the stable volatiles that can be generated are the products of methylation or trimethylsilylation reactions.

The limitation of the technique lies in the lack of selectivity of the detectors available, and therefore exhaustive purification has to be

Figure 3.1 Types of immunoassay.
(*a*) *Radioimmunoassay* (RIA): radiolabelled and 'cold' antigen compete for a limited number of antibody-binding sites. The reagents in the assay are in solution, and separation of the 'bound' and 'free' phases is achieved by either precipitation of the antibody or adsorption of the 'free' antigen.
(*b*) *Enzyme-linked immunosorbent assay (ELISA)*:
    (i) Direct method: the technique uses the antigen linked to an enzyme rather than radioactivity. The antibody is bound to a solid phase such as the well of a microtitre plate, and 'free' and enzyme-linked antigen molecules compete for the immobilized binding sites. At equilibrium, the 'free' phase is decanted and the quantity of 'bound' enzyme determined after the addition of the enzyme's substrate. Most commonly, the antigen is linked to either alkaline phosphatase or horseradish peroxidase, since these enzymes exhibit high activity against substrates which produce products which are coloured or fluorescent and are therefore readily quantifiable.
    (ii) Indirect method: this method employs the conjugation of the antigen to a protein which is immobilized to the walls of a support such as the well of a microtitre plate. 'Free' antigen and antibody are added to the reaction vessel and the antibody molecules bind to either the immobilized or the 'free' antigen. The soluble antibody-antigen conjugate is decanted away. An enzyme-linked second antibody which specifically recognizes the antiserum in which the primary antibody was raised is introduced into the reaction vessel. This secondary antibody binds to the immobilized conjugate. After the liquid phase has been removed, the substrate of the enzyme linked to the secondary antibody is added and the amount of product quantified. Although this method is potentially more time-consuming than the direct method, the use of a second antibody improves the sensitivity of the assay.
Abbreviations: Ab, antibody; *Ag*, antigen; *Ag**, radio-labelled antigen; *Ab-Ag*, antibody-antigen conjugate; *Ab-Ag**, antibody-radiolabelled antigen conjugate; *Ag-E*, enzyme-linked antigen; *S*, substrate; *P*, product; $_B$-Ag, protein-antigen conjugate; $_2Ab$-*E*, enzyme-labelled second antibody. (After Robins, 1986.)

carried out prior to injection into a GC. The extent of purification can be reduced by choice of capillary columns and packing material, coupled with shrewd manipulation of column temperature. For quantification, retention time and peak area are related to those of an injected standard. While some would argue that these are insufficient parameters on which to base the identification of a compound, flame ionization detection has been traditionally accepted as an appropriate method by which to quantify ethylene. In principle, however, any molecule which ionizes on combustion in a hydrogen flame would give a positive reading in this system. The electron capture detector is more selective since it only registers compounds that are electron-negative. ABA naturally has this property; however, other PGRs can be detected if the derivatization process generates a linked halogen group.

Other detectors for use with GC systems have been recently developed. These include an alkali flame ionization detector which specifically monitors nitrogen- or phosphorus-containing molecules, and a photoionization detector for ethylene (Figure 3.2). The high sensitivity of these detectors (Table 3.2) makes them a useful adjunct to the GC armoury for PGR analysis.

3.3.5 Gas chromatography–mass spectrometry

The selectivity problem of GC analysis may be resolved by linking a mass spectrometer to the system as a functional detector (Heddon, 1986). The mass spectrometer analyses the fragments released from ionized compounds on the basis of their mass-to-charge ratio. Every

Table 3.2 Sensitivity of detectors to plant growth regulators

Detector[a]	Sensitivity[b]	Plant growth regulator
Electron capture	High	ABA
Flame ionization	Low/medium (>10nl 1^{-1})	Ethylene
Fluorimeter	Medium/high	IAA/GA/CK/ABA
Mass spectrometer	Low/medium	IAA/GA/CK/ABA
Photoionization	Medium/high (<10nl 1^{-1})	Ethylene
Refractometer	Low	IAA/GA/CK/ABA
Ultraviolet	Medium	IAA/CK/ABA

[a] As a comparison, the sensitivity of immunoassays to IAA/GA/CK/ABA is high
[b] Low sensitivity — detection of more than 100ng per sample
  High sensitivity — detection of less than 100pg per sample

For further details see Reeve and Crozier (1980).

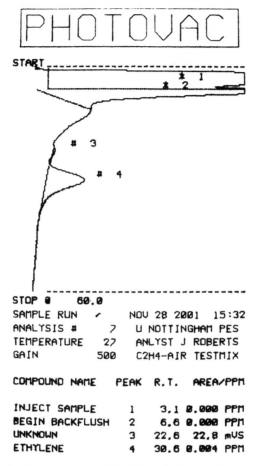

Figure 3.2 Analysis of gas sample containing 4nl $^{-1}$ ethylene by 'Photovac' portable gas chromatograph with a photoionization detector. Printout 150% natural size; separation is achieved with a 4' Carbopack-BHT 100 column at ambient temperature.

compound produces a characteristic mass spectrum from which it can be identified by comparison with spectra from known standards. This technique has been used very effectively for the identification of trace amounts of PGRs, and in particular the different groups of GAs.

Although a mass spectrometer will record all the major ion fragments generated during the ionization process, it can be selectively tuned for single ions. In this mode, specific ion fragments indicating particular compounds of interest can be monitored. This powerful technique, called selective ion monitoring (SIM), allows the rapid identification and quantification of specific molecules from what would otherwise be a forest of peaks on the GC trace (Figure 3.3). The reliability of SIM is based on the 'uniqueness' of the ion fragments chosen for scrutinization. Further confidence in the analysis can be obtained by monitoring more than one specific ion and the ratio between them. This enhanced selectivity is obtained at the expense of reduced sensitivity, as the machine spends less time focused on one particular ion.

3.3.6 High performance liquid chromatography

Quantification of all the major groups of PGRs apart from ethylene can be achieved using HPLC. This is a powerful technique, and can utilize all the conventional chromatographic separation methods, including phase partition, reversed phase partition, adsorption and ion exchange systems. HPLC is more effective than conventional chromatography because solvent is forced at high pressure through a column containing material of fine particle size. Samples are therefore eluted quickly and with good resolution.

The advance of HPLC has been largely hampered by the detection methods currently available. The most commonly employed is the UV spectrometer or refractometer. IAA, CKs and ABA can all be detected with a UV spectrometer; however, GAs will only absorb UV light at around 210–230 nm. This restricts the solvents that can be used in the separation process, since both ethyl acetate and acetic acid absorb at these wavelengths. A further method which has been employed with success for PGR analysis is electrochemical detection. A coulometric type of electrochemical detector has been used to assist the studies of Wright and Doherty (1985). These workers have found that the detection limit of this technique is approximately 2.5 pg of IAA, and that this resolution allows estimations to be made of the auxin content of single half-nodes from stems of *Avena fatua*. The sensitivity of this method suggests that coulometric detection could become the method of choice for IAA measurement in the future.

Fluorescence detectors are also finding favour in conjunction with HPLC systems. This technique requires the generation of fluorescent derivatives of PGRs but is more sensitive than the UV detector, and has been used successfully to assay for IAA and GAs. Laser excitation offers further possibilities in conjunction with fluorimetric detection

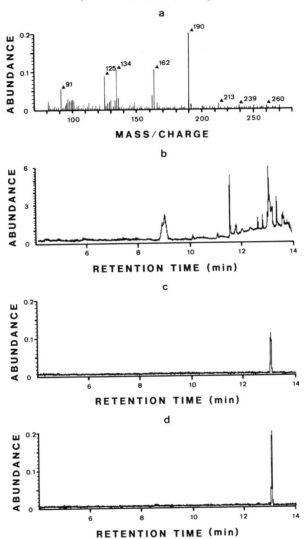

Figure 3.3 (*a*) Mass spectrum of authentic *cis*-methyl ABA. Note major ions at 162 and 190. (*b*) Total ion count of GCMS scan of plant extract. (*c*), (*d*) Selective ion monitoring of extract at 162 and 190 respectively. GC condition: column BP₁ 25m × 0.22mm (internal diameter), helium carrier gas at 1ml min⁻¹, initial temperature 135°C, then programmed to 250°C at 10°C min⁻¹.

systems. Prototype instruments that have been constructed using micro-bore columns suggest that by using a laser source, fluorescence detector limits for IAA may be in the low femtogram range. Indeed, it has been suggested that such a system is potentially capable of detecting individual molecules of this PGR.

3.4 The state of the art

The analysis of plant tissue extracts has now reached the level of sophistication where picogram quantities of PGRs may be routinely detected and their identity confirmed with a high degree of precision (Table 3.3). In an attempt to ascertain the probabilities of identification being accurate, information theory has been applied to certain analyses (Reeve and Crozier, 1980). Whilst this contribution to the field may be intellectually stimulating it does not get to grips with the real question of whether endogenous auxins, GAs, CKs, ABA or ethylene regulate plant differentiation and development. Indeed, at this stage it is appropriate to consider which of the approaches outlined in this chapter is most likely to generate answers to this question.

A cell is composed of a number of compartments, the complexities of

Table 3.3 Analytical methods for plant growth regulators

Plant growth regulator	Method of analysis	Problems associated with analysis
Auxins	GC-MS HPLC-F Immunoassay	Chemical and enzymic oxidation
Gibberellins	GC-MS Immunoassay	Ring arrangements oxidation
Cytokinins	GC-MS HPLC-UV Immunoassay	Adsorption to glassware, enzymic degradation during extraction
Abscisic acid	GC-EC GC-MS Immunoassay	Isomerization to *trans*, *trans*-configuration
Ethylene	GC-FID GC-PID	Contamination by other volatiles from enclosing vessels

Detectors: EC, electron capture; FID, flame ionization; F, fluorescence; MS, mass spectrometer; PID, photoionization; UV, ultraviolet.

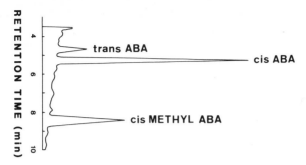

Figure 3.4 Reverse phase HPLC separation of *cis* and *trans* ABA isomers. Separation achieved using a 5μm Spherisorb column 250mm × 4.6mm (internal diameter); solvents, methanol/water/formic acid, 50:45:5; flow rate 1ml min^{-1}; UV detector at 252nm.

which can only be visualized with the aid of techniques such as high voltage electron microscopy. The number of PGR molecules in any one compartment is dependent upon such factors as rates of import and export, and sites of synthesis and degradation. Within each compartment, only a fraction of the molecules may be available for activity, the others being sequestered in some bound form. A further variable is that the compartments may not all have equal significance in eliciting a developmental event, since some may be spatially separated from the site of action of the PGR. It is clear from this simple appraisal that subtle changes in cellular PGR levels could occur, yet remain undetected by some of the crude approaches that are currently adopted. For instance, it is clearly inappropriate to treat plant extracts as if they were from tissues composed of a homogeneous population of cells containing equal numbers of PGR molecules. Therefore, as increasingly sensitive methods of detection are developed, they should be employed to assay PGR levels in specific tissues rather than to increase the accuracy of measurement in whole organs. An example of the benefit of this approach has come from studies on GA levels in pea seeds. Effort has been concentrated on determining the level of GAs in developing seeds because this tissue is a rich source of these PGRs. Many different GAs have been found, but their functions remain obscure. Since it is now possible to handle meristematic tissues, attention has been turned to quantifying GAs in tissues of the pea embryo. It turns out that this tissue contains low levels of GA$_1$ and only a few other GAs. The data from this study are far more interesting than the identification of the numerous GAs which accumulate in seeds, because these few low-level GAs are functionally important, as are their interconversions to inactive forms (see Chapter 6). In the future, with the aid of techniques such as laser-induced fluorescence, it may be possible to determine even the low levels of PGRs within individual cells.

The generation of antibodies to PGRs offers potential probes for cellular location studies. By tagging antibodies with markers such as fluorescent or electron-dense materials, the cellular and subcellular distribution of PGRs may be visualized. This is an important aspect of PGR analysis, since some developmental events could be regulated by a modification in import or export of a PGR rather than a gross change at a cellular level. In addition, other subtle controls may exist such as the conversion of inactive to active PGRs (or *vice versa*) at discrete subcellular locations adjacent to the 'sites of action'. This scenario highlights the further consideration that studies on PGR levels and localizations must be reinforced by knowledge of the location and activities of PGR receptors (see Chapter 9).

The techniques of analysis described, such as HPLC or GCMS, are powerful ones for the separation of compounds in a plant extract. Primarily these methods have been used to quantify and identify known compounds. An alternative approach is to use these systems to scan plant extracts for potential regulatory molecules. This is not necessarily a 'needle in a haystack' approach, since by shrewd choice of developmental event or access to isogenic mutants exhibiting an aberrant phenotype, it may be possible to spotlight compounds of significance. This technique has already made a significant contribution to our understanding of the biosynthetic pathway of ABA, and preliminary reports indicate that in this manner a putative floral inhibitor has been isolated (see Chapter 5).

To summarize, significant progress has been made over the last 10 years in the development of techniques by which to analyse plant extracts for PGRs. The methods currently available are sensitive and reliable, and most plant tissues can be analysed with ease. The use of antibodies as analytical tools should prove to be an exciting development in this field, and with the aid of these probes it should be possible to monitor the intracellular concentrations of PGRs and their subcellular locations. Clearly, studies such as these will make a dramatic contribution to our understanding of the role of PGRs in plant growth and development.

CHAPTER FOUR

HORMONES AND THE CONCEPT OF SENSITIVITY — A RATIONAL APPROACH

From the foregoing chapters, it is evident that the bioassay has been an important tool for both the isolation and quantification of PGRs. However, its 'success' has relied on the principle that the biological response of a plant cell is proportional to the amount of PGR that is applied. This assumption is, perhaps, an oversimplification of the truth.

The response elicited by a PGR can be considered to be the result of an interaction between two components (Figure 4.1). The first is the number of 'active' PGR molecules adjacent to the cellular site of their action. This value is dependent on the biosynthesis and metabolism of the PGR, and also the velocity of its import and export. The second is the 'sensitivity' of the cell to the endogenous PGR. This property is ill-defined, but can be envisaged to include changes in PGR receptor number, or affinity, or in the events subsequent to receptor occupancy. This chapter examines some of the factors which may influence endogenous PGR concentrations and the sensitivity of cells to PGRs, and attempts to rationalize the present conflict over which, if either, of these two elements may play the predominant role in the regulation of plant development.

4.1 The hormone concept

The concept that plant growth and development is regulated by levels of endogenous PGRs originates from the experiments carried out by Charles and Frances Darwin on phototropism in grass coleoptiles (see Chapter 2). Their observations indicated that although the coleoptile apex was the site of perception of light, the growth response which resulted in bending was located in cells further down the tissue. They therefore reached the conclusion that phototropic bending was regulated by the translocation of a chemical messenger (later identified as IAA), which moved from the apex to the elongating zone of the coleoptile where it induced differential growth. Plant physiologists of the time were quick to see the similarity between this idea and the concept of 'hormones', a term coined by the animal physiologist Ernest Starling in

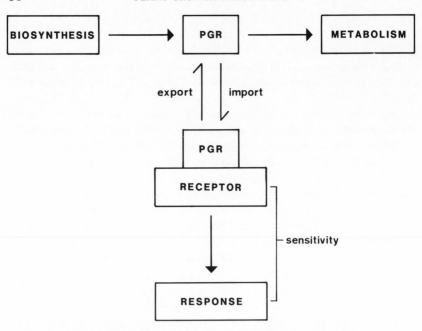

Figure 4.1 Factors which may affect the impact of a PGR on a developmental event. The response elicited is dependent on both the number of PGR molecules which are available to bind with the receptor and the sensitivity of the cell to the PGR. The sensitivity of the cell may be governed by such factors as receptor number, receptor affinity, or events that follow receptor occupancy.

1905 to describe regulatory chemicals which were produced in one organ and transported in the bloodstream to another where they acted. It was a short step from here to the 'christening' of auxins, GAs, CKs, ABA and ethylene as plant hormones. The accuracy of this classification remains to be seen.

4.1.1 Sites and regulation of biosynthesis

A fundamental tenet of the definition of a chemical as a hormone is that its biosynthesis should be restricted to a discrete group of cells. Although there is no evidence that PGRs can strictly fulfil this criterion, there is good reason to believe that not all cells and tissues have an equal capacity to synthesize auxins, GAs, CKs, ABA and ethylene.

Two main approaches have been adopted to localize the sites of biosynthesis of a particular PGR, and both of these have significant drawbacks. In the first method, radiolabelled PGR precursors are fed to plant tissues and potential sites of conversion screened for the presence of radiolabelled PGR. The success of this approach relies on an accurate knowledge of the biosynthetic pathway of the PGR, and an ability to introduce biosynthetic precursors into the appropriate cellular compartment for them to be converted. In the second method, PGR levels are monitored in specific organs or tissues throughout development. A change in PGR level during ontogeny is then assumed to reflect a change in biosynthetic capacity. An example of the potential fallacy of this assumption is the conclusion based on Went's bioassay data that the maize coleoptile tip is a major biosynthetic site of auxin. There is now wide support for the hypothesis that the majority of the auxin in the apex of maize coleoptiles originates from the endosperm. Maize endosperm tissue contains myo-inositol IAA, and this conjugated auxin is transported to the coleoptile tip where it is converted to IAA (see Chapter 2).

A third approach which has been adopted seeks to correlate biosynthetic intermediates and the enzymes responsible for their conversion with particular plant tissues. This method has been used to localize sites of biosynthesis of IAA, and although the pathway remains open for debate, an examination of the distribution of putative intermediates has led to the proposal that the major sites of biosynthesis of this auxin are meristematic tissues, rapidly growing buds and developing seeds. The most convincing demonstrations of auxin biosynthesis by plant cells have come from studies of tissue cultures where cells can be studied in isolation from other tissues of the plant. The tissue culture system may open up avenues by which to explore the compartmentalization of PGR biosynthesis. However, the extent to which such systems may mimic the situation in a whole plant is unknown.

The enzymes and intermediates of GA biosynthesis are clearly defined for some tissues, and this allows a more rigorous study of the sites of PGR synthesis to be carried out. In the pea plant, levels of GAs are highest in developing seeds, and cell-free systems have been isolated from pea embryos which contain all the enzymes necessary for the conversion of mevalonic acid to GA_{51} and GA_{29}. These observations confirm the view that immature seeds can synthesize GAs. Cell-free enzyme systems capable of incorporating ^{14}C-mevalonic acid into *ent*-kaurene have also been used to study sites of GA biosynthesis. The *ent*-kaurene-synthesizing system has been found in preparations from expanding leaves, internodes, shoot tips, petioles and stipules from both tall and dwarf pea plants, but is absent in root tips. Although interpretation of these results must be made cautiously, there is a good correlation between rates of *ent*-kaurene biosynthesis in the extracts and growth of the tissues, which would give credence to the hypothesis that the rates are an accurate reflection of GA biosynthesis. Reports have appeared implicating a role for plastids in GA biosynthesis, and some of the enzymes involved in the conversion of kaurenoids to GAs have been found in these organelles. However, attempts to incorporate mevalonate into GAs using preparations of isolated chloroplasts have, as yet, proved unsuccessful.

Reputedly, the major site of CK biosynthesis in higher plants is the root and in particular the root tip. This hypothesis has been formulated largely on the basis of the demonstrations that high CK levels can be found in xylem sap as well as in cultures of rapidly growing excised root tips. Since rootless tobacco shoots grown from callus tissue can synthesize the CK i^6Ade from $(8 - {}^{14}C)$-adenine, shoot tissues must also have the capacity to synthesize CKs. Other sites which have been proposed include the developing seed and the embryonic axis. Greatest advances in our understanding of CK biosynthesis have come from studies on crown gall tissues of *Vinca rosea* and tobacco. Although these tissues provide a useful model system on which to probe PGR biosynthesis, their relevance to the situation in the intact plant has yet to be determined.

Sites of ABA biosynthesis are widespread throughout the plant. As a result of water stress, ABA levels have been reported to increase in isolated leaves, stems, buds, roots and root tips. In addition, ABA has been shown to accumulate in fruits and seeds. Reports that isolated epidermal layers cannot synthesize ABA have yet to be substantiated, particularly in the light of the demonstration that guard cell protoplasts can respond to osmotic stress by synthesizing ABA (Weiler *et al.*, 1982). The observation that as much as 90% of the ABA in a leaf is located in mesophyll chloroplasts has led to the proposal that ABA is synthesized there. However, this hypothesis is very much open to question, since the chemical properties of the ABA molecule ensure that, irrespective of the subcellular site of biosynthesis, ABA would become sequestered within the chloroplast (Hartung *et al.*, 1981). Once again this highlights the potentially misleading conclusions that can be reached by correlating the presence of a PGR with its biosynthetic site.

Of all the PGRs, perhaps sites of ethylene biosynthesis can be ascribed to a tissue with greatest confidence. This is because the biosynthetic pathway is well documented, and its intermediates and enzymic components are easy to assay. In addition, putative sites of biosynthesis can be probed with the aid of chemicals such as aminoethoxyvinylglycine (AVG) which specifically block ethylene biosynthesis. The evidence strongly suggests that, although all plant cells at some stage in their life cycle may have the capacity to synthesize ethylene, the level of production can vary dramatically. For instance, during the ripening of climacteric fruit, ethylene production may rise by a factor of over one hundred (see Chapter 8).

If our knowledge of the cellular sites of PGR biosynthesis is considered fragmentary, then that relating to intracellular sites is virtually non-existent. Progress may be made in this latter area with the aid of cell cultures and *in-vitro* systems; however, extrapolations of the results from such isolated cell systems to the intact plant will have to be made cautiously. Further advances await the development of techniques such as those described in the previous chapter to determine the intracellular localization of PGRs and the key enzymes involved in their biosynthesis.

If the rate of PGR biosynthesis by plant tissues can alter, it is important to consider the mechanisms which might regulate this change. However, in the absence of a complete understanding of the enzymes involved in PGR biosynthesis, it is very difficult to make accurate assessments of biosynthetic rates. Clearly, monitoring the level of a PGR is not an adequate measure of this parameter.

Convincing information on the regulation of PGR biosynthesis is available only for ethylene. This is because the potentially rate-limiting members of the biosynthetic pathway of this PGR, such as the enzyme ACC synthase and the ethylene-forming enzyme system, have been identified and can be readily assayed. The activity of these enzymes is stimulated during developmental events such as ripening and senescence (see Chapter 8), and in both these processes the increase in ethylene biosynthesis is thought to be regulated by the gas itself. Wounding also stimulates ethylene biosynthesis, and an enhancement of ACC synthase activity has been detected within 20 minutes of cell damage (Figure 4.2). Application of many of the other PGRs to plant tissues have been reported to elevate ethylene production. In particular, treatment of shoots, leaves, fruits, flowers and roots with auxin can significantly stimulate ACC synthase activity and hence ethylene production within a few hours of treatment (Figure 4.3). This is a good example of

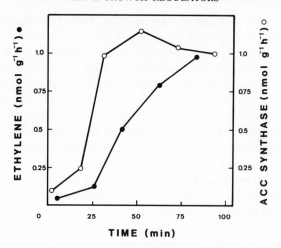

Figure 4.2 Time-course of ethylene production and ACC synthase activity after wounding green tomato fruit pericarp tissue. (Redrawn from Boller and Kende, 1980.)

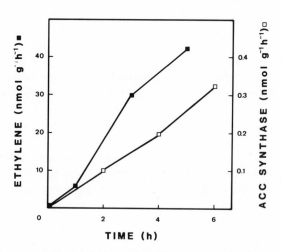

Figure 4.3 Time-course of ethylene production and ACC synthase activity after application of IAA (5×10^{-4}M) to mung bean hypocotyl tissue. (Redrawn from Yoshii and Imaseki, 1982.)

how PGRs can interact, and highlights the drawbacks of studying PGRs in isolation from each other.

From the discussions in Chapter 2 it is evident that PGR-deficient mutants are proving to be of great value in the identification of intermediates in the biosynthetic pathways of these compounds. Furthermore, it is expected that such mutants will ultimately assist in the establishment of the enzymes responsible for the regulation of the pathways. Purification of the rate-limiting enzymes in PGR biosynthesis, and identification of the mRNAs coding for these enzymes, is an important step towards probing the regulation of PGR biosynthesis. Work is currently in progress to carry out this procedure on enzymes such as ACC synthase and the 2β- and 3β-GA-hydroxylases, and it is hoped that this will be achieved in the near future.

4.1.2 Transport and its regulation

The second part of the definition of a hormone is that the compound should be transported from its site of production to its site of action. Therefore, if PGRs are correctly classified as plant hormones, it would be predicted that they should be readily transported around a plant. Furthermore, it might be expected that the intercellular movement of PGRs would exhibit signs of coordination.

Auxins, GAs, CKs, and ABA can be detected in the xylem and phloem of a range of species, demonstrating that PGRs can circulate within the plant's vascular system. However, this information tells us little about the physiological significance of such movement. Detailed studies of PGR transport have been undertaken primarily on excised sections of plant material. The conventional technique is to sandwich segments of shoot or root tissues between blocks of agar, one of which is used as a donor of radiolabelled PGR and the other as a receiver. The donor blocks are routinely loaded with a synthetic PGR analogue to reduce the contribution that PGR metabolism makes during the course of the experiment. The velocity of PGR transport is determined by measuring the rate of accumulation of radioactivity into the receiver. The validity of this crude technique is open to question, since data from experiments of this kind do not always compare favourably with those obtained from studies on intact plants.

Using the excised segment technique, the basipetal polarity of auxin transport, first detected in etiolated *Avena* coleoptiles by Went in 1928, has been confirmed as a general feature of shoot tissues. Transport of

IAA is saturable, occurs at a rate of approximately $10-15$ mm h^{-1} depending on the tissue, and is specifically inhibited by compounds such as triiodobenzoic acid (TIBA) and naphthylphthalamic acid (NPA). These chemicals have therefore been routinely employed to probe the role of auxin transport in a particular developmental event. The balance of evidence from studies on root segments indicates that IAA transport in this tissue is predominantly acropetal. Therefore a flux of auxin is maintained throughout the plant from shoot to root apex.

No clear picture has emerged from investigations into the transport of other PGRs. Whilst there are instances of tissues exhibiting polarity of GA, CK or ABA movement, in general they are not convincing and are commonly obtained from experiments where high concentrations of PGR have been loaded into the donor block. The gaseous nature of ethylene makes studies of its translocation within plants difficult; however, investigations have been made of the movement of its immediate biosynthetic precursor ACC. Experiments carried out on waterlogged tomato plants have demonstrated that ACC is mobile within the xylem sap, and moves from the anaerobic environment of the roots to the shoot, where it is converted to ethylene (Bradford and Yang, 1980).

Polar transport of IAA in both shoot and root tissues is maintained against a concentration gradient and is inhibited by anaerobic conditions. These properties of the system indicate that polar transport has a requirement for metabolic energy, and this might be consumed by events related to either the uptake or efflux processes. Because of the difficulties of examining these components in a multicellular tissue, progress in elucidating the mechanism of IAA transport has come primarily from work on populations of single cells. Investigations carried out by Rubery and Sheldrake (1974) on crown gall cell suspension cultures of *Parthenocissus*, and by Raven (1975) on the giant alga *Hydrodictyon*, led to the formulation of the 'chemiosmotic polar diffusion hypothesis' of IAA transport (Goldsmith, 1977). This simple theory is centred on two basic properties of the IAA molecule. Firstly, undissociated molecules of the auxin are hydrophobic and therefore readily traverse lipid membranes. Secondly, since IAA is a weak acid, its rate of dissociation to form the lipophobic anion IAA$^-$ increases with rising pH. From a consideration of these facts, it can be predicted that plant cells surrounded by an acidic environment, such as the cell wall, would readily take up IAA. However, once inside the cell, the auxin would dissociate in the more basic environment of the cytoplasm and the IAA$^-$ anions would become effectively trapped. If an IAA$^-$ carrier was localized within the plasma membrane, this would reduce anion accumulation by catalysing its efflux down a concentration gradient. Polarity of auxin transport would be achieved by the maintenance of an asymmetric distribution of efflux sites within the cell, and ATP would be required to sustain the pH and electrical gradients necessary to drive the transport process (Figure 4.4).

Since the hypothesis was proposed, auxin efflux carriers have been detected in a variety of tissues which exhibit polar transport. Furthermore, kinetic studies of IAA movement in *Cucurbita pepo* hypocotyl

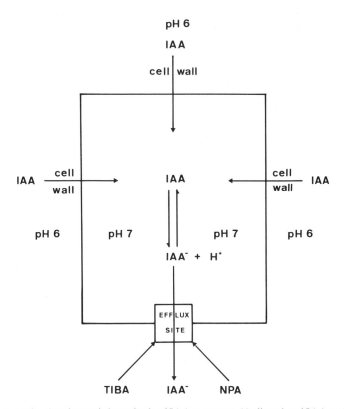

Figure 4.4 The chemiosmotic hypothesis of IAA transport. Undissociated IAA molecules are hydrophobic and therefore readily enter the cell. In the more alkaline environment of the cytoplasm, the auxin dissociates into the lipophobic anion IAA⁻. Efflux of this anion is mediated by a carrier located at the basal end of the cell. Both triiodobenzoic acid (TIBA) and naphthylthalamic acid (NPA) specifically bind to the efflux carrier and render it inoperative.

segments indicate that the efflux carrier is specifically blocked by TIBA and NPA, although these chemicals appear to bind to a different site on the carrier than IAA (see Chapter 9). The NPA binding protein has been purified and monoclonal antibodies used to identify its cellular location in pea stems by immunofluorescence techniques (Figure 4.5). In accord with the predictions made by the chemiosmotic hypothesis, NPA-binding proteins are specifically associated with the basal end of cells, and these are situated adjacent to the vascular tissue (Jacobs and Short, 1986). NPA is not a natural constituent of plant cells, and there-

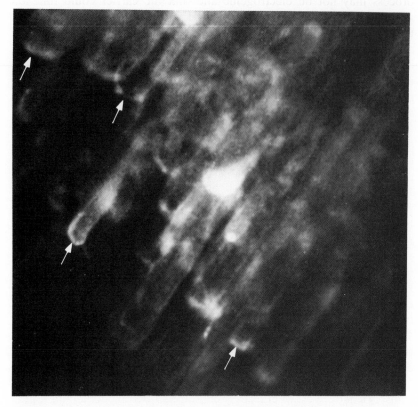

Figure 4.5 Localization of the presumptive auxin efflux sites in pea stem tissue. The sites were visualized using indirect immunofluorescence techniques with the aid of monoclonal antibodies raised against NPA binding proteins. Fluorescence is clearly associated with the basal regions of stem cells. (Photograph courtesy of Dr Mark Jacobs, and reproduced with permission from Jacobs and Gilbert (1983). Copyright © 1983 by the AAAS.)

fore a question that has been frequently posed is the nature of the molecule which binds to the NPA receptor *in vivo*. It has been reported recently (Rubery and Jacobs, unpubl.) that flavanoids such as quercetin can displace bound NPA from plant tissues, and therefore it is plausible that certain phenolics could play a role in regulating IAA transport in plant tissues.

The development of an *in-vitro* IAA transport system using membrane vesicles from *Zucchini* hypocotyls has provided an opportunity to examine the transport components of the plasmamembrane (Hertel *et al.*, 1983). These vesicle studies have established that the efflux process may not be electrogenic and that the carrier could function as a symport, transporting IAA^- and H^+ across the plasmamembrane. Furthermore, the studies have revealed the existence of a specific auxin uptake mechanism. Although the chemiosmotic hypothesis predicts that influx of IAA will occur by diffusion, an auxin uptake carrier may be necessary to augment the process in order to sustain rapid IAA transport. The uptake carrier has the capacity to translocate IAA^- and H^+, and could be either electrogenic or electroneutral in character.

A specific uptake carrier has also been identified for ABA, and this has been located in roots and suspension-cultured cells of *Phaseolus coccineus*. Characterization of the carrier has indicated that it is a symport, transporting both ABA^- and H^+ to maintain electroneutrality. Intriguingly, carrier activity is restricted to the elongating region of the primary root (Milborrow and Rubery, 1985). Whether this observation has any significance for the role of ABA in the regulation of root growth remains to be seen.

Plants have clearly evolved a highly effective translocation system for the cell-to-cell movement of auxins, and therefore in many ways this PGR could be considered to fulfil the second criterion of the hormone concept. However, as yet there is little evidence that the intercellular transport of GAs, CKs or ABA is regulated in a coordinated manner. This may be because the cells which synthesize these PGRs are also responsive to them, or that they are only separated from them by a short distance. Alternatively, it may be sufficient to manipulate GA, CK, ABA or ethylene (ACC) movement by regulating phloem and xylem transport. However, of crucial significance to our understanding of the activity of PGRs is knowledge of the movement of these compounds from one cellular compartment to another. Our information on this aspect of PGR transport is rudimentary. Major advances in this area will require the generation of further *in-vitro* systems for the study of PGR uptake and efflux, and the development of immunological procedures to probe the subcellular distribution of PGRs (see Chapter 3).

If PGRs do function as plant hormones, then any stimulus which enhances or inhibits movement of these compounds could have a significant impact on plant growth and development. In particular, the highly polarized auxin transport system is a potential target for manipulation by a range of stimuli. Generally, environmental factors have not been found to alter the polarity of IAA transport, although exceptions to this rule have been recorded. For instance, it has been convincingly demonstrated that the polarity of IAA movement in the leaf sheath base of the grass *Echinochloa colonum* is dictated by the direction of gravity (Wright, 1981). Furthermore, there have been numerous reports that gravitropic or phototropic stimulation of shoot tissue can promote lateral transport of IAA, although this remains a subject of great debate (see Chapter 7). It should be remembered, however, that an environmental stimulus could have a significant impact on the velocity of IAA transport without altering the overall polarity of the tissue. Few detailed studies have been undertaken to examine this, although those that have been carried out suggest that both light and gravity can influence the velocity of auxin movement.

If excised tissue segments are allowed to age prior to experimentation, then the capacity of the material to transport auxin declines. A profile of the distribution of radiolabelled auxin within the tissue under these conditions shows that whilst uptake from the donor block takes place, auxin accumulates in cells adjacent to the cut surface and is not transported. Pretreatment of segments with kinetin maintains the ability of the tissue to transport auxin, which suggests that the reduction in transport capacity may be correlated with the onset of senescence. Certainly, as petiole tissues senesce *in vivo* their capacity to transport auxin has been shown to decline. Polar auxin transport is also inhibited by exposure of plant tissues to ethylene (Figure 4.6) and this phenomenon may make an important contribution to this gaseous PGR's ability to promote abscission.

Little information is available on the regulation of movement of other PGRs. It has been hypothesized that water deficit may reduce transport of CKs from the roots to the shoots and that this, in conjunction with elevated ABA levels, may precipitate stomatal closure (Davies *et al.*, 1987). If this hypothesis proves correct, then the regulation of movement of PGRs within the phloem and xylem could make a significant contribution to plant development.

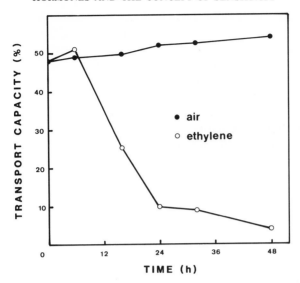

Figure 4.6 Effect of ethylene ($25 \mu l\ l^{-1}$) on the basipetal transport of ^{14}C-IAA in midrib sections of citrus leaves. Following transport, the sections were cut transversely into three pieces of equal length and the radioactivity determined in each piece. The transport capacity of the section is defined as the percentage of total uptake transported below the apical segment. (Redrawn from Riov and Goren, 1979.)

4.1.3 Target cells

The final component of the concept is that a hormone should act at a specific target site. In animals, the target tissues occupy specific locations within the body and hence are clearly recognizable. The meristematic arrangement of plants' apices does not generate discrete 'organ-like' structures, and there has been considerable debate over the existence of target cells. It has, however, been argued that the aleurone layer could be considered as a target tissue for GAs (see Chapter 6), and that abscission zones may be composed of ethylene-responsive target cells (see Chapter 8). In addition, it has been proposed that other putative target cells can be identified on the basis of their growth response to IAA or ethylene (Osborne, 1982). If specific target cells for PGRs do exist, then they should be distinguishable from their neighbours by molecular determinants. An immunological approach has been adopted in an attempt to highlight such molecular determinants. Antibodies have been raised against proteins extracted from putative target tissues (ethylene-responsive abscission zones) and competed with antigens from neighbouring non-target tissue. The rationale behind this approach is that the competition reaction should eliminate common determinants between the two types of cells and thus pinpoint any determinants which are unique to the ethylene-responsive cells (Osborne, *et al.*, 1985). The preliminary results of this interesting study indicate that abscission zone cells do have unique antigenic determinants and that these can be employed to specifically stain abscission zone cells using immuno-gold labelling techniques. This work is still in its infancy, and a considerable further research effort will be necessary to determine the chemical identity of any specific determinants which are discovered.

4.2 The concept of sensitivity

Much of the research undertaken on PGRs has been based on the premise that the endogenous levels of these compounds dictates the magnitude of a developmental response. Indeed, the very isolation of PGRs through the utilization of the bioassay relied on this assumption. Recently this concept has been challenged, and it has been cogently argued that the sensitivity of plant tissues to a PGR is a critical factor regulating development. This hypothesis is not new, but Trewavas, its modern-day protagonist, has suggested that its significance has become forgotten in the planning of recent research strategies.

The thrust of Trewavas's argument rests on two assertions (Trewavas, 1982): firstly, that PGRs fail to fulfil a strict hormonal role, and secondly, that many developmental events are preceded not by a change in PGR levels, but by a modulation in PGR sensitivity. Trewavas argues that if PGR levels directly controlled development, then a fluctuation in the endogenous concentration of these molecules should, in theory, precede or parallel the progress of a developmental event, whereas in practice such demonstrations are rare. In addition, an examination of the dose–response curves to applied PGRs reveals that they commonly span over four orders of magnitude (Nissen, 1985), and he considers that this feature would be unacceptable if PGR concentration finely tuned the progress of development.

The views of Trewavas have been greeted with considerable criticism. This is because he has adopted the stance that sensitivity is the overriding criterion regulating plant development, and has thus sought to devalue a contribution made by changes in PGR concentration. Ironically, in an effort to undermine traditional ideologies, re-introduction of the concept of sensitivity affords an opportunity to retain specific dogmas of PGR involvement in development when correlations with endogenous levels of these compounds are absent.

As it stands, the term 'sensitivity' is too imprecise. In Trewavas's view, sensitivity is dictated by the number of functional receptors that a cell has in its armoury. However, there are many other ways in which the sensitivity of a cell to an applied PGR might be modulated. For instance, the 'uptake efficiency' or 'metabolic capacity' of the transport pathway could change during ontogeny, and this would moderate the response of a tissue towards an exogenous PGR. Under these circumstances, the apparent sensitivity of the system would be directly proportional to the concentration of the PGR at its active site within a cell. If the change in sensitivity originates from a fluctuation in receptor number, this could be considered as a modification in 'receptivity', while if the basis of the change was due to the capacity of a receptor to bind the PGR, this could be described as an attenuation in 'affinity'. Sensitivity changes arising from the sequence of events subsequent to receptor occupancy could be considered to represent adjustments in 'response capacity'. These terms, which have largely been proposed by Firn (1986), are useful since they strive to challenge research workers to pinpoint the origin of cellular sensitivity. Furthermore, the basis of these terms can be used to model theoretical dose–response curves representing the different sensitivity scenarios (Figure 4.7). Interestingly, the conclusion from this modelling exercise is that whilst the varying mechanisms which could account for sensitivity do alter the impact of a PGR on a developmental process, the response invariably remains dependent on PGR concentration.

4.3 Hormones and the concept of sensitivity — a rational approach

The sessile nature of plants has forced them to adopt a flexible developmental programme that can be co-ordinated by environmental cues. It is evident that PGRs are potential candidates to transduce these environmental signals; however, a crucial question is whether biochemical and physiological events are triggered by a change in PGR levels, or whether PGRs act to co-ordinate the progression of a developmental sequence after it has been initiated. The current state of thinking on this issue has become polarized. On the one side, there are the traditionalists who favour the hypothesis that PGRs are plant hormones and as such are the limiting factors in plant development. On the other are those who support the concept that developmental control is achieved through

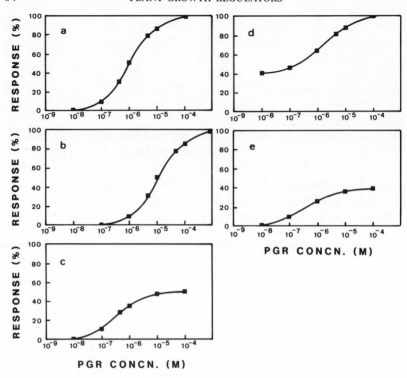

Figure 4.7 Dose–response curves modelled from theoretical data quantifying the impact of a PGR on a developmental response. The curves have been obtained by altering certain parameters underlying the relationship between the PGR and the developmental process. (*a*) Basic dose–response curve. (*b*) Response curve generated by reducing the receptor affinity or transport capacity by a factor of 10, or increasing the metabolic capacity by a factor of 10. (*c*) Curve generated by halving the response capacity of the tissue. (*d*) Response curve generated by a tissue which already has a high endogenous concentration of the appropriate PGR. (*e*) Response curve originating from a tissue which has a reduced receptivity to the PGR. For further details see text. (Redrawn from Firn, 1986.)

modulations in the sensitivity of cells to PGRs. In an attempt to rationalize this conflict it is pertinent to examine each of these viewpoints in turn.

A critical examination of the evidence has revealed that PGRs do have some features in common with putative plant hormones. For instance, although it is evident that PGR biosynthesis is not restricted to discrete tissues, it is also clear that some plant cells have a greater capacity to synthesize a particular PGR than others; moreover, the biosynthetic capacity of a cell may vary during its ontogeny. Secondly, at least in the case of IAA, a highly regulated polar transport mechanism has evolved which could be important in targeting the distribution of this PGR. There is no evidence that transport of the other PGRs is regulated; however, in general cells may both synthesize a particular PGR and respond to it, and as Pinfield points out, you don't need a postal service if you write a letter to yourself! Thirdly, circumstantial evidence has recently appeared which supports the assertion that target cells for PGRs may exist. However, if certain cells are targets for a specific PGR, then other cells are not, and as such must be deemed insensitive. Thus within the traditionalist framework of PGRs as hormones, the concept of cellular sensitivity must also be accepted.

There can be little doubt that plant tissues do exhibit differential sensitivity to PGRs, as judged by their response to application of these compounds, and throughout the following chapters numerous references will be made to this phenomenon. However, regardless of its innate sensitivity, a cell will remain 'quiescent' in the absence of an appropriate effector molecule. Therefore, the evocation of sensitivity to a PGR is restricted by the presence of that compound, and as discussed in the previous section, the degree of response is in part dependent on the level of the PGR. This train of thought leads us to the logical position that control exerted via PGR levels or through cellular sensitivity need not be considered as mutually exclusive hypotheses. Rather, they can be seen as aspects of the same mechanism of developmental regulation.

This rationalization can be examined further by considering a developmental system where PGR levels and cellular sensitivity may both make a significant contribution. Potato tubers normally remain in a dormant state throughout the season of growth until some weeks after harvest. The duration of this period of innate dormancy is largely dictated by the cultivar and the temperature of storage. A detailed study of tuber growth has revealed that dormancy can be overcome by application of CKs. Bud growth can be initiated by CKs immediately after tuberization and towards the end of a 12-week storage period at 10°C. Between these two stages, there is a window of insensitivity to applied CKs. Quantification of endogenous CKs by radioimmunoassay has shown that levels of this PGR are high immediately after tuberization, and then fall rapidly prior to storage. Throughout storage of tubers at 10°C, the level of CKs is low, but begins to rise prior to the termination of innate dormancy. It is clear from this study that potato tuber

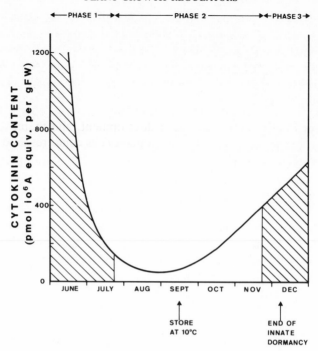

Figure 4.8 Phases of sensitivity to cytokinins and cytokinin levels in potato tubers during growth or storage at 10°C. During Phase 1, tubers are dormant, even though they can be induced to sprout by application of CKs, and CK levels are high. In Phase 2, tubers are insensitive to CKs and endogenous levels of this PGR are low. During phase 3, CK levels increase in concert with the return of sensitivity to applied CKs and the termination of dormancy. (After Turnbull and Hanke, 1985.)

growth passes through three distinct phases (Figure 4.8). During phase 1, tubers are dormant, even though they are sensitive to applied CKs, and endogenous levels of this PGR are high. On entering into phase 2, tubers become insensitive to applied CKs and endogenous levels are low, and finally, during phase 3, CK levels increase concomitantly with an increase in sensitivity and the termination of dormancy. This detailed study supports the contention that a developmental programme such as dormancy may be regulated temporally by both PGR levels and sensitivity. Moreover, the paradoxical nature of phase 1 highlights the present inadequacies in our quantification of PGR levels and also our assessment of cell sensitivity.

In this chapter, we have assessed the mechanisms which regulate

PGR biosynthesis and metabolism, and the sites where this occurs. We have also considered the inter- and intracellular transport of PGRs and how these processes could attenuate the levels of these compounds at their active sites within plant cells. The modulations of such levels may directly dictate the developmental response of a cell, or development might be controlled through changes in sensitivity to PGRs. The balance of evidence indicates that both of these parameters could play a significant role in the regulation of development, and thereby provide a highly flexible framework by which plant cells could respond to their environment.

CHAPTER FIVE

CELLULAR DIFFERENTIATION AND MORPHOGENESIS

One of the major aims of this book is to appraise critically the role that PGRs play in plant development. Certain developmental phenomena which have been reported to be influenced by PGRs, such as juvenility, vernalization and the photoperiodic induction of flowering, involve the transition of shoot apices from one state of differentiation to another. These observations raise the possibility that PGRs could have a role to play in regulating the differentiation of plant cells *per se*. The process of differentiation is of fundamental significance to our understanding of plant growth and development, and yet studies in this area have been rare. This chapter examines the role that PGRs may play in influencing apical transitions, and considers the mechanisms which could regulate cellular differentiation and morphogenesis.

5.1 Juvenility

It is a common phenomenon for seedlings of herbaceous and woody species to pass through a 'juvenile' stage in their development (Wareing, 1987). The most characteristic feature of juvenile plants is their inability to flower under conducive conditions; however, other differences in form and behaviour may also distinguish them from adults of the same species. For instance, juvenile and adult phases of ivy (*Hedera*) can be distinguished by leaf shape and size, phyllotaxy, pigmentation and rooting capacity.

The duration of the period of juvenility varies considerably from one species to another. In herbaceous plants it is rarely more than a few weeks in length, whereas in woody plants it can persist for many years. The primary prerequisite for the transition from juvenile to adult is related to plant size, and this state must be reached regardless of the time period over which it is attained. Once transition to the adult phase has taken place, the condition is stable throughout subsequent mitotic divisions, until it is reversed by events associated with meiosis. In *Hedera, Citrus* and *Prunus*, rejuvenation can also be induced by repeated applications of GA_3. Observations such as these have prompted the suggestion that PGRs might play a role in regulating the transition from the juvenile to adult phase. The evidence in favour of this hypothesis remains circumstantial, and attempts to induce maturity in juvenile plants by the application of PGRs or inhibitors of PGR biosynthesis have been largely unsuccessful. Premature maturity has been induced successfully in juvenile ivy plants by water stressing or chilling the roots. Therefore some signal from the root tissues could play a role in regulating the onset of the phase change in the shoot apical meristem.

The molecular basis of the phase change from juvenile to adult is unknown. The stability of the process indicates that it is an inductive phenomenon, and that it involves differential gene expression. Whether PGRs play a role in the promotion or inhibition of phase transition *in vivo* remains open to debate. However, with the aid of the techniques of molecular biology it is now appropriate for answers to this question to be sought.

5.2 Flowering

The transition from a vegetative to a floral apex is an event of strategic importance in the growth and development of a plant. In addition, the process has significant commercial interest to both the agricultural and horticultural industries. Considerable attention has therefore been focused on the mechanism of floral induction and evocation. This section highlights recent progress that has been made in our understanding of the regulation of floral differentiation, and assesses the role that PGRs play in the flowering process.

5.2.1 Photoperiodism

A major environmental signal which regulates the differentiation of floral structures is light and in particular daylength. Not all plants require a specific photoperiod in order to flower, but those that do have been broadly classified as 'short-day plants' (SDPs) or 'long-day plants' (LDPs) on the basis of whether they require the daylength to be shorter or longer than a critical duration in order to initiate flowering. Not only is the length of the light period important in photoperiodism, but so too is the duration of the dark period, and it is evident that both SDPs and LDPs have endogenous rhythms which underlie the photoperiodic response. A discussion of the mechanism of time measurement is beyond the scope of this book but interested readers are recommended to the review by Thomas and Vince-Prue (1984) for further details.

By exposing different parts of an SDP or LDP to inductive photoperiods it has been possible to localize the site of perception of the photoperiodic stimulus. It is clear that, although it is the apex which ultimately differentiates to generate floral structures, it is the leaves that perceive daylength. It follows, therefore, that photoperiodism relies on the transmission of some 'signal' from the leaves to the apex of a plant. Confirmation of this has come from the demonstration that leaves from an induced SDP will stimulate flowering when grafted on to uninduced plants. The first successful grafting experiment of this kind was carried out by Chailakhyan in 1936 on tobacco. The results of this study led him to the conclusion that the photoperiodic induction of flowering was mediated by a graft-transmissible promoter, and he coined the term 'florigen' to represent the active principle. Since then, despite intensive research, efforts to isolate and identify florigen have met with no success. One possible explanation for their failure could be that floral induction involves the removal of inhibitors rather than the production of a promoter. Indeed, flowering might be initiated by a reduction in the level of an inhibitor which reduces the ability of leaf tissues to divert specific nutrients away from the apex (Evans, 1987). Certainly, a substantial body of evidence has now accumulated which invokes a role for floral inhibitors in the photoperiodic response. For instance, some SDPs and LDPs, such as *Chenopodium amaranticolor* and *Hyoscyamus niger* respectively, can be made to flower under non-inductive photoperiods if all their leaves are removed. Floral induction of leafless *Hyoscyamus niger* plants can be prevented if a single leaf is grafted back on to the defoliated plant. Furthermore, in many experiments where photoperiodically induced leaves are grafted on to uninduced plants, flowering is only initiated if the recipient is first stripped of its foliage. Recently, a putative floral inhibitor has been isolated from cotyledons of the SDP *Pharbitis nil* (Jaffe *et al.*, 1987). This compound has been identified as *bis*(2-ethylhexyl)hexane dioate (BEHD), and has been shown to have the capacity to inhibit flowering of SDPs, LDPs and day-neutral plants. If this compound is indeed an endogenous floral inhibitor, its discovery represents a significant scientific breakthrough and could pave the way for the elucidation of the photoperiodic regulation of flowering.

There have been a number of documented reports that application of auxins, CKs, ABA or ethylene can influence the flowering process (see Schwabe, 1987). Indeed, one of the commercial uses of ethylene-generating chemicals is to induce flowering in bromeliads (see Chapter 11). However, the effects of auxins, CKs, ABA or ethylene on the regulation of flowering are limited to only a few species, and thus it is unlikely that any one of them is the elusive florigen. In contrast, there is a substantial amount of evidence that GAs could play a role in the photoperiodic regulation of flowering. For instance, exposure of a range of LDPs to GAs will stimulate flowering under non-inductive photoperiods, inductive treatments may change both the quantity of endogenous GAs and their identity, and inhibitors of GA biosynthesis can inhibit flowering of some LDPs maintained under inductive photoperiods. The issue is complicated by the fact that, in general, the stimulatory effects of GA are restricted to rosette plants. Flower initiation in such plants is preceded by stem elongation and this raises the possibility that the impact of applied GAs may be on stem growth rather than floral induction *per se*. Moreover, this hypothesis is supported by the demonstration that the GA biosynthesis inhibitor AMO 1618 can suppress bolting in the LDPs *Silene armeria* and spinach without preventing the plants from flowering under inductive photoperiods.

It would appear from these studies that none of the known PGRs is florigen. However, circumstantial evidence implicates GAs in the flowering process in some plants. Whether GAs are required for the synthesis or action of a floral promoter, or alternatively promote flowering as an indirect effect of their capacity to stimulate stem elongation growth, is presently unknown. A major difficulty which has been encountered by all those who have sought to isolate floral promoters or inhibitors is the development of a suitable flowering bioassay, since many of those that have been adopted have proved unreproducible and time-consuming. One possible exception to this is the utilization of thin cell-layer explants from tobacco floral branches which have revealed that certain oligosaccharides at very low concentrations can induce either flowers or vegetative buds depending on the pH of the medium (Tran Thanh Van *et al.*, 1985). However, what is really needed is a rapid assay based on the earliest changes which may be induced in the vegetative apex after receiving the floral stimulus. Work is currently in progress (Francis, 1987) to elucidate some of the changes that take place at a cellular level at the time of floral evocation, and these studies may lead to the development of a suitable bioassay with which to go in search of endogenous floral regulators with renewed enthusiasm. The characterization of flowering-related cDNAs may also prove useful, particularly if they can be shown to be specific to flower induction. Until then, we may have to rely on more speculative approaches such as that adopted by Jaffe in his isolation of BEHD.

5.2.2 Vernalization

Environmental conditions other than light can have a profound effect on the differentiation of floral structures. In particular, some plants require a period of low temperatures before flowering can be induced. This phenomenon is known as vernalization. Most commonly the cold treatment is given to young seedlings, although in certain species the imbibed seed is also receptive. In *Brassica* species, floral initials are induced during the vernalization period, but in many plants they appear only after transference to higher temperatures and often in response to precise photoperiodic regimes.

Although it is more difficult to identify the site of perception of low temperature than that of light, localized cooling temperatures have been applied to plants which require a vernalization stimulus. In general, only the stem tip need be exposed to low temperatures for the vernalization treatment to be effective, although it is possible that all dividing cells are potential sites of perception of a cold stimulus. Indeed, in chicory, the root system has the capacity to induce flowering, since an unvernalized shoot will flower after being grafted on to a vernalized root. Although this observation implies that a transmissible stimulus could be involved in the vernalization process, grafting experiments on other vernalization-sensitive species have met with only limited success. Clearly there are some similarities between photoperiodic induction of flowering and vernalization. However, the crucial question is whether low temperature stimulates the production of a 'florigen-like' compound (vernalin), or whether the treatment sensitizes specific cells at the apex so that they are able to respond to a floral stimulus.

The observation that application of GA_3 can replace the chilling requirement in some species has led to the proposal that this PGR might play a key role in the vernalization process. Further support for this hypothesis has come from the demonstration that vernalized plants may contain higher endogenous GA levels than unvernalized plants. However, these results must be interpreted with caution, since as in photoperiodism, those species that are induced to flower by GA treatment are essentially restricted to rosette plants. In these species, low temperatures not only stimulate the differentiation of the floral apex but also shoot elongation. It is possible therefore, that the capacity of GAs to induce flowering in unvernalized plants may be once again a secondary effect of their ability to stimulate stem elongation. Further work is necessary to investigate this possibility, and to determine whether the process of vernalization generates floral promoters *de novo*, or sensitizes the apex to the reception of a floral signal.

5.3 Sex expression

Once the appropriate signal has been received by the cells of the apical meristem, the programme of events leading to floral development is set in motion. In monoecious and dioecious species, the floral signal may be interpreted to induce the differentiation of either staminate or pistillate flowers. The ratio of male to female flowers is largely dependent on environmental conditions such as temperature and the duration of the photoperiod; however, it can also be manipulated by application of PGRs.

The role of PGRs in the regulation of sex expression in monoecious species has been most extensively investigated in cucumber (*Cucumis sativus*). In general, if plants of this species are exposed to ethylene the differentiation of pistillate flowers is stimulated, whereas treatment with the ethylene antagonist silver nitrate induces the formation of staminate flowers. Auxins can also act as feminizing agents; however, since these compounds have the capacity to stimulate ethylene biosynthesis, their influence may be indirect. GAs have a converse effect to auxin and ethylene, since they promote the differentiation of male flowers; furthermore, the differentiation of male flowers can be suppressed by the application of inhibitors of GA biosynthesis (Pharis and King, 1985). There is no evidence that the effects of GAs are mediated through a change in endogenous ethylene levels, and the influence of these two PGRs appears to be unrelated. In other monoecious species, the effect of PGRs on sex expression may not conform to the same pattern, and in *Zea mays*, for instance, it has been demonstrated that application of GA_3 to 3–4-week-old seedlings stimulates the differentiation of female flowers. Intriguingly, later applications of this PGR can cause sterility or even promote masculinization.

PGRs have also been reported to regulate the differentiation of floral structures in dioecious plants. For example, treatment of young seedlings of spinach with GA_3 has been shown to enhance the formation of female flowers, whereas exposure of similar plants to GA biosynthesis inhibitors has the converse effect. However, other studies on spinach have produced conflicting results, which implies that PGRs may not be the only factors which can have a significant impact on the regulation of sex expression.

It is clear from the examples described above that PGR applications can manipulate sex expression in some species. However, these observations provide no evidence that PGRs regulate the differentiation of floral structures *in vivo*. For example, an exogenously applied PGR could modify the form of a monoecious plant and in this way alter the number of sites at which flowers of a particular sex could develop. As a result, the PGR would change the ratio of male to female flowers without having a direct effect on floral differentiation. Further work is therefore necessary to determine how changes in endogenous PGR levels influence floral sex expression; this will require the use of specific agents to regulate PGR biosynthesis, and will be facilitated by the generation and characterization of PGR-deficient mutants. The monitoring of endogenous PGR levels will provide additional evidence. However, it is clear from the discussions in Chapter 3 that a detailed analysis of those cells which comprise the floral primordia may be necessary, and this will necessitate the further development of highly sensitive methods of PGR detection.

5.4 Vascular differentiation

An examination of a transverse section through the stem of any vascular plant reveals the degree to which differentiation of plant tissues is strictly co-ordinated. This pattern is most clearly exemplified by the symmetrical distribution of xylem and phloem cells which comprise the vascular tissue. The mechanism of pattern formation in plants is unknown, although in animals there is evidence that it relies on the movement of a specific stimulus which provides the differentiating cells with information about their position with respect to other cells within the tissue. There is good reason to believe that the positional differentiation of vascular tissue is also co-ordinated by a transmissible stimulus. For instance, if apical leaf primordia are excised, further differentiation of vascular bundles at the apex fails to take place, but if a developing leaf is grafted back on to the apex, vascular differentiation is reinstated. In addition, vascular differentiation can be induced in undifferentiated callus tissue by grafting a young bud on to the top of the callus. In both the grafting experiments described, the differentiation of vascular tissue progresses basipetally from the site of the graft; therefore it is likely that the stimulus which induces vascular differentiation originates from the young leaf tissues.

The ability of leaf primordia to induce vascular differentiation can be mimicked by auxins (see Jacobs, 1984). Localized application of IAA to callus tissue can induce the differentiation of not only xylem and phloem, but also cambial cells, especially if the PGR is applied in conjunction with sucrose. Moreover, the auxin will induce the formation of vascular tissue in the cortex of pea stems when applied to decapitated seedlings. Since young leaves are thought to be a major source of IAA (see Chapter 4), it has been proposed that this auxin is a natural evocator of vascular differentiation. Further evidence in support of this hypothesis has come from experiments on the regeneration of vascular tissue following wounding. When a vascular bundle within a section of *Coleus* stem is severed by a transverse cut, continuity of the strand is restored by regeneration of vascular tissue within the pith. Regeneration progresses basipetally from the upper severed end and is dependent on the presence of an intact leaf positioned above the wound. If the leaf is delaminated, regeneration does not take place unless IAA is applied to the debladed petiole. This series of observations suggests that IAA may play an important role in inducing vascular differentiation in pith tissue during the wound response.

If IAA is a natural regulator of vascular differentiation, the polarity of transport of this PGR (see Chapter 4) would ensure that the basipetal development of the vascular tissue is maintained. Indeed, it was reported as long ago as 1892 by Vochting that the direction of vascular differentiation is dictated by tissue polarity, and it has since been shown, in grafting experiments where one of the members is inverted, that regeneration follows the original polarity of the tissues even though this may prevent the establishment of new root and shoot connections (Figure 5.1). If the tissue is wounded extensively, then the natural polarity of auxin flow can be prevented. Under these circumstances it has been suggested that IAA accumulates at the site of wounding until a new direction of flow is initiated by diffusion. This renewed flow may not only induce vascular differentiation but also impose a new polarity of auxin transport on the tissue, and as a consequence generate a positive feedback between polarity and auxin movement (Sachs, 1986).

In some tissues, xylem differentiation is influenced by CKs. This is not mediated through an effect of the PGR on cell division, since it has been demonstrated that the induction of xylogenesis in isolated mesophyll cells of *Zinnia elegans* is dependent on the presence of both auxin

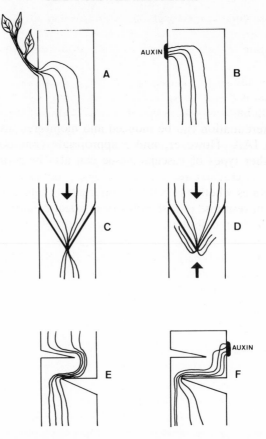

Figure 5.1 Vascular differentiation in plant tissues. New vessels (wavy lines) can be induced by grafting on a lateral bud (*A*), or by the localized application of auxin (*B*), to a decapitated stem. Differentiation of vessels is dependent on tissue polarity. New vascular tissue follows the original polarity in tissue grafts (*C*) even if the stock is inverted (*D*). (Arrows point to the original direction of the roots.) New vessels will readily differentiate to circumvent a transverse wound (*E*) and this phenomenon can be mimicked in a decapitated plant by the application of auxin. (Redrawn from Sachs, 1986.)

and CKs and takes place in the absence of mitosis. The interaction between auxin and CKs in the regulation of differentiation warrants further study, as it has been known for over 30 years, from the classic experiments of Skoog and Miller, that the ratio of these two PGRs can dictate the formation of root or shoot meristems in tissue cultures. There are conflicting opinions over a role for GAs in xylogenesis since some tissue culture systems respond favourably to the addition of the PGR while others do not. One plausible explanation for this is that the concentration of GA and the time of its application are critical (Pearce *et al.*, 1987).

Much of the work carried out on the role of auxin in vascular differentiation is based on the induction of xylem vessels in plant tissues. This system has been commonly adopted because of the ease with which vessel differentiation can be induced and monitored after exposure of tissues to IAA. However, under appropriate conditions, the induction of other types of vascular tissue can also be promoted by IAA treatment. A crucial question which remains unanswered, therefore, is 'What dictates whether a cell will differentiate into xylem, phloem or cambium in response to a stimulus from a leaf primordia or to IAA?'. This is a fundamental question in the study of differentiation and poses a major challenge to plant molecular biologists, biochemists and physiologists which has yet to be grasped. In the following section we discuss the possible mechanisms responsible for regulating the positional differentiation of plant cells.

5.5 Morphogenesis

In the previous section we have considered the ways by which differentiation of vascular tissue may be coordinated. However, this is only one of a host of examples where the genetic potential of undifferentiated plant cells is expressed in a controlled manner. In the absence of such control, the specific arrangement of cell types which characterize different plant species would not exist. The mechanism of morphogenesis is therefore of fundamental significance to our understanding of the regulation of plant growth and development.

Much of the work on morphogenesis has been undertaken on animal systems, and from these studies two general principles have emerged (Wolpert, 1981). Firstly, some mechanism must exist which can supply a cell with the information it needs to determine its position relative to defined boundary regions. Secondly, the cell must be able to interpret this signal and as a result differentiate along a specific pathway (Figure 5.2). This interpretation step is quite distinct from that which supplies the positional information. The application of these rules to plant morphogenesis generates some intriguing possibilities. For instance, the positional information which traverses shoot meristems could be interpreted in a different way from that which crosses root meristems. More-

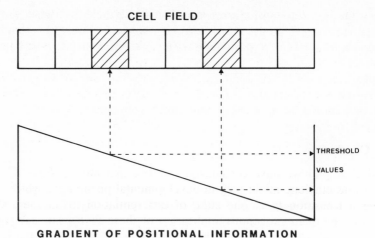

Figure 5.2 The concept of positional information. Differentiation of cells is proposed to occur in response to the concentration of one (or more) morphogen(s). The morphogen gradient assigns a positional value to each cell within the field and each cell interprets this value and differentiates in response to it. Since the interpretation step is quite distinct from that which supplies the positional information, two cells which occupy different positions within the gradient could differentiate in the same way. (Redrawn from Wolpert, 1981.)

over, the response of cells within a vegetative apex might be different from those within a floral apex.

In one of the few papers on plant morphogenesis, Holder (1979) has made the novel suggestion that PGRs might have the capacity to modify the way that plant cells interpret their positional cues. If this hypothesis was correct it would be predicted that PGRs would be able to alter the pattern of cell differentiation in root and shoot meristems in a predefined way. It is tempting to speculate that the regulation of vascular differentiation by IAA is one example of this phenomenon, and that the manipulation of floral evocation and sex expression by GA and ethylene may be others. It is conceivable in these instances that the PGRs could be supplying the tissues with the positional information itself, although, as Holder argues, if this was true one might expect to see much grosser distortions in cellular patterns.

The concept that pattern formation is the result of a cell interpreting its positional information provides a unifying hypothesis to explain aspects of plant development. For instance, it raises the possibility that florigen, anti-florigen or vernalin could be stimuli which manipulate the interpretation of positional cues, and that the juvenile-to-mature phase transition which occurs in some species could be co-ordinated by a similar mechanism. If this hypothesis is correct, in order to understand how plant differentiation is regulated we need to go in search of both the signals that provide positional cues and those that dictate how this information is interpreted. Clearly, this is a very challenging task, but also one that should prove highly rewarding in the future.

5.6 Conclusions

In this chapter we have considered the part that auxins, GAs, CKs, ABA and ethylene might play in developmental phenomena which involve a transition from one state of differentiation to another. On balance, although exogenous application of these PGRs can influence processes such as maturation, flowering, and sex expression, there is little experimental evidence to indicate that they play a significant role in the regulation of these events *in vivo*. On the other hand, there is convincing evidence that auxin and CKs can have an effect on the process of differentiation itself. Clearly, auxins can stimulate xylogenesis in several *in-vitro* systems, but in addition, IAA would appear to be of strategic importance in the maintenance of vascular continuity in intact plants. The mechanism by which IAA influences differentiation is unknown, but it is conceivable that it causes cells to interpret information about their position in a new way. If this hypothesis is correct, it raises the possibility that this might be a general feature of PGRs, and could account for the impact of these compounds on some of the developmental processes described previously. As yet, there is no experimental evidence to support this novel scheme. However, perhaps the time is right to explore its implications for the manipulation of plant growth and development.

CHAPTER SIX

SEED DEVELOPMENT, DORMANCY AND GERMINATION

Developing seeds are a rich source of auxins, GAs, CKs and ABA, for, along with storage proteins, carbohydrates and lipids, they accumulate PGRs. The abundant quantities of PGRs in seeds have made them the subject of many extraction, identification and quantification experiments on which those seeking to unravel PGR chemistry have been able to 'cut their teeth' and refine their analytical techniques. From these studies have sprung investigations into the role of PGRs during seed development. The aim of this work has been to determine if PGRs are simply deposited in seed as a potential carbon source to be consumed during germination, or alternatively regulate seed development by conveying signals from the mother plant. Thus PGRs laid down in the developing seed might regulate its growth and development once it has matured and been shed, controlling such phenomena as dormancy and the growth which takes place during the first few hours of germination. This would appear to be a sensible strategy for such a critical and precarious stage in a plant's life. In this chapter we examine the evidence that PGRs play such a strategic role.

6.1 Seed development

Gibberellins have been found in seeds of a wide range of plant species and at considerably higher levels than those reported in other tissues. In general, seed GAs are of low biological activity compared with the high activities of GA_1 and GA_3 found in vegetative tissues. While GAs may be transported to the developing seed, it is clear that growing seeds can and do synthesize and metabolize GAs. There is some evidence to suggest that GAs are important in sustaining early embryo growth, but most studies have concentrated on the role of GAs during later stages of seed development. Since there is a great diversity in the chemistry and biological activity of different GAs (see Chapter 2), it is important to know (i) the absolute amounts of different GAs in a tissue at each stage of development, (ii) the biological activity of each of the GAs involved, and (iii) their rates of metabolism.

GA metabolism has been most elegantly and rigorously studied in developing *Pisum sativum* seeds, which contain 9 different GAs (Sponsel, 1985). During pea seed growth, there are two peaks of GA activity, the first coinciding with fruit set and early embryo growth. During the rapid growth phase, biologically active GA_9 and GA_{20} predominate and are correlated with starch accumulation. These GAs are inactivated during seed maturation by 2β-hydroxylation to GA_{51} and GA_{29} respectively, and may be subject to tissue-specific catabolism (see Chapter 2). It is not yet clear whether the high levels of GAs that are present in developing seeds during the period of highest growth rate are a consequence or cause of growth. However, reducing the levels of GAs with GA biosynthesis inhibitors has little effect on seed yield, suggesting that the levels of GAs in seeds are more than sufficient for growth. Certainly the regulation of embryo or cotyledonary growth may require only minute quantities of active GAs at specific sites within these organs.

Cytokinins are also abundant in developing seeds (Van Staden, 1983) and are predominantly in the form of N^6-substituted derivatives. In addition to the 'free' CKs described in Table 6.1, a number of bound forms occur in *Zea mays* as glucosides. CKs present in seed may be imported from root and shoot tissues. Certainly radiolabelled zeatin

Table 6.1 Cytokinins in developing seeds and fruit

Cytokinin	Species	Seed state
Zeatin	*Zea mays*	Immature kernel
	Cucurbita pepo	Immature seed
	Prunus cerasus	Developing fruit
	Cocos nucifera	Liquid endosperm
	Pyrus malus	Immature seed
	Gossypium hirsutum	Ovules
Dihydrozeatin	*Lupinus luteus*	Immature seed
	Gossypium hirsutum	Ovules
Trans-ribosylzeatin	*Zea mays*	Immature kernel
	Cocos nucifera	Liquid endosperm
	Pyrus malus	Immature seed
	Gossypium hirsutum	Ovules
Cis-ribosylzeatin	*Zea mays*	Immature kernel
	Triticum aestivum	Mature embryo
Isopentenyladenine	*Gossypium hirsutum*	Ovules
Isopentenyladenosine	*Gossypium hirsutum*	Ovules

applied to the leaves of lupin accumulates within the seed, and removal
of fruit and seed results in elevated CK levels in the remainder of the
plant. However, seed tissues may also synthesize CKs. For instance, pea
seeds accumulate CKs when excised pods are cultured *in vitro*, although
the CKs may originate from either the pod, or seed, or both. Whatever
their source, seed CKs can be correlated with particular developmental events. In general, CK levels are high during periods of rapid cell
growth and division either of the embryo or storage organs, and these
levels decline as the seed matures, possibly as a result of metabolism to
O-glucosides (Figure 6.1).

Developing cereal grain accumulates auxin in the endosperm, and in
wheat this reaches a peak shortly before maximum grain fresh weight is
attained. In dicotyledons, endogenous auxin levels increase in concert
with the phases of most active seed cell growth. In some seeds, biphasic
growth patterns of endosperm cells are reflected by two auxin maxima.
As seeds mature, IAA is conjugated to bound forms with myo-inositol,
arabinose and other carbohydrates.

A consistent pattern of PGR levels is apparent from the above dis-

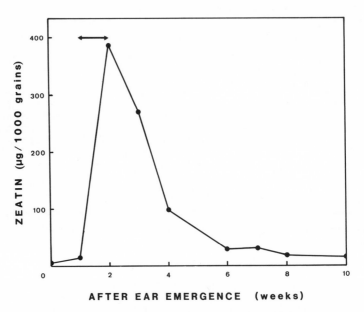

AFTER EAR EMERGENCE (weeks)

Figure 6.1 Change in cytokinin content of wheat grains during their development.
Horizontal bar designates the period of anthesis. (Redrawn from Bewley and Black,
1982.)

cussions. As seeds mature, 'free' PGRs are converted to bound forms. This may reflect a careful packaging process for subsequent use during germination, or alternatively it may simply be a convenient form of PGR disposal. As yet we do not know whether peaks of active PGRs during seed development are causally or consequentially related to cell growth. The use of PGR-deficient or sensitivity mutants may help to clarify this issue.

Our understanding of the role of ABA during seed development is, encouragingly, a little further advanced. ABA increases during the development of seed of a wide range of species, and reaches a maximum at approximately the same time as does seed dry weight. At maturation, once again there is a substantial decline as ABA is metabolized to phaseic and dihydrophaseic acids. The role of ABA during seed development may be twofold. When immature embryos of *Brassica napus* are isolated and cultured *in vitro*, their ABA levels drop and they germinate precociously. Immature embryos of wheat and many other plant species behave similarly. Application of ABA prevents this germination. Further support for the notion that seed ABA may play a role in the prevention of precocious germination is the demonstration that some viviparous maize mutants possess low levels of ABA in the kernels. Furthermore, vivipary can be induced in some plants by treatment with compounds such as fluridone, which have been demonstrated to reduce endogenous ABA levels. However, it is worth remembering that osmotica are as effective as ABA in the prevention of germination of isolated immature embryos, and therefore precocious germination might be regulated by water availability. Evidence that ABA is involved in normal embryogeny is more convincing (see Quatrano, 1986). If immature embryos of *Brassica napus*, and several other species that are synthesizing storage proteins in the intact seed, are isolated, levels of storage protein mRNAs decline rapidly. The levels can be restored by the addition of ABA. Furthermore, the expression of a spectrum of storage protein and related genes can be induced by ABA in isolated embryos of wheat, at a time well before the genes are expressed in the intact seed. This and other evidence suggest that ABA has the capacity to promote maturation-specific morphological and biochemical events in immature embryos.

6.2 Dormancy

Dormancy can be considered to begin at an early stage in seed development, for, as described in the previous section, immature embryos can 'germinate' when isolated from the seeds but are prevented from doing so *in vivo*. Commonly this phase of development terminates concurrently with seed maturation, and the mature seeds will germinate when provided with a suitable environment. Sometimes dormancy is not maintained through to seed maturation, and in this case, under moist conditions, seeds can germinate on the mother plant: examples are viviparous mutants of corn and the preharvest sprouting of wheat. Alternatively, dormancy may extend beyond the time of seed maturation, and under these circumstances mature seeds will not germinate even when provided with a 'conducive' environment, as in wild oat. Seeds remain in this state for a finite time which is dependent on the type and depth of dormancy. This phase of seed development clearly constitutes an efficient strategy for the spread and survival of the species.

In view of the efficacy of ABA in suppressing immature and mature embryo growth, it is tempting to speculate that this PGR plays a role in dormancy *in vivo*. One theory of dormancy is that it is controlled by a balance between endogenous inhibitors and promotors of germination. Application of GAs, CKs or ethylene can overcome dormancy in some species, and therefore it is not unreasonable to theorize that the duration of dormancy is the summation of the interaction between these PGRs and ABA. In practice, this hypothesis is too simplistic, and PGRs may play only a minor role in the regulation of dormancy.

If ABA is responsible for preventing germination *in vivo*, then it might be expected that more ABA would be present in dormant than in non-dormant seed, and ABA levels would decline in dormant seed prior to the onset of germination. It should be stressed at this point, however, that the mechanism may involve events more subtle than a gross change in ABA levels. For example, tissue or subcellular redistribution of the PGR may constitute a far more precise trigger for germination. Alternatively, changes in the ability of the embryo to perceive and respond to PGRs could be equally effective. In some cases, such as in embryos of dormant and non-dormant cultivars of peach, there is a good correlation between endogenous ABA levels and dormancy. Similarly, seed of ABA-deficient mutants of *Arabidopsis thaliana* are non-dormant, while, in contrast, seeds of wild-type plants of the same species are dormant (Figure 6.2) (Karssen *et al.*, 1987). In other species, such as *Avena fatua*, levels of ABA are the same in both dormant and non-dormant seed. Some seeds, such as those of *Lactuca sativa* cv. Great Lakes and *Acer saccharinum*, contain considerable quantities of ABA yet are non-dormant. These observations have been used as an argument against a

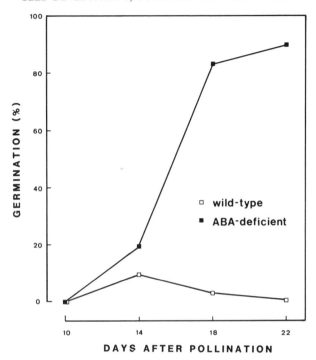

Figure 6.2 Precocious germination of seed isolated from wild-type or ABA-deficient
Arabidopsis thaliana genotypes. Seeds were removed from the siliquae at the indicated
time after pollination and germination was recorded after 7 days. (Redrawn from Karssen
et al., 1987.)

role for ABA in dormancy, although in these species, since the cellular
location of the PGR is unknown, it is conceivable that the embryo is
never 'exposed' to the ABA which can be extracted from the seed.

In seeds such as apple and oak, dormancy can be broken by chilling.
In these species, levels of ABA have been reported to decline in seed
stored both at 5°C and at 20°C, even though the latter temperature
maintains the dormant state. After several weeks at these temperatures,
ABA may reach undetectable levels in both those seeds which remain
dormant and those which do not. These various data do not provide
compelling evidence in favour of a role for ABA in the regulation of
dormancy, in spite of the fact that minute quantities of this PGR will
effectively induce a state closely resembling dormancy in a range of
different species.

D

GAs are particularly effective at promoting seed germination, irrespective of whether the origin of dormancy is the seed coat or the embryo itself. CKs are generally less effective than GAs, even though they can convincingly counteract the effects of ABA. Ethylene can also overcome dormancy in some seed. Because these three PGRs are found in seeds, it has been proposed that together they may control the release from dormancy.

In some dormant seed the chilling requirement can be replaced by GA or CK. Does chilling elevate GA or CK levels *in vivo*? In general, there have been no consistent demonstrations that chilling treatment precipitates an increase in the levels of either GAs or CKs in a pattern convincingly related to dormancy breakage. However, many measurements of levels of these PGRs have been carried out using bioassays alone, and there is a good case for re-evaluating PGR levels using modern methods of analysis. One study that has employed such techniques has shown that chilling of hazel seed results in a slight increase in GAs which rise more dramatically when the seed is transferred to higher temperature. It would appear that chilling enhances the capacity of the hazel embryo to synthesize GA_1 and GA_9, thus enhancing GA levels in germinating seed. Dormancy in hazel can be broken by application of GAs.

The involvement of ethylene in dormancy breaking is well established, and appears to involve those seeds with coat-imposed dormancy, such as cocklebur. In this species the upper seed is more dormant than the lower one, and during after-ripening the lower seeds exhibit an elevated capacity to synthesize ethylene. Synthesis of ethylene precedes germination, and application of the gas early in seed imbibition stimulates germination. These observations suggest that ethylene may play a role in dormancy regulation in this species.

6.3 Germination

It is evident from the preceding section that PGRs, and in particular GA and ABA can have a significant impact on germination. Before it is possible to assess their role *in vivo* it is necessary to consider those events which constitute germination.

The onset of germination has been widely considered to be marked by signs of radicle emergence. Some workers have questioned this notion, since the earliest molecular events which initiate germination will undoubtedly be invisible to the naked eye. Radicle emergence may perhaps therefore be best considered as the first *visible* sign of germination. In some species, radicle protrusion is mediated by cell elongation, and cell division occurs later. There is convincing evidence that elongation growth is modulated by GAs and auxin (see Chapter 7), and it is tempting to speculate that visible germination may involve these two PGRs. The fact that GAs can promote germination, and that seed from GA-deficient *Arabidopsis thaliana* mutants will not germinate (Karssen *et al.*, 1987) may be evidence to support this (Figure 6.3). Interestingly, the fungal toxin fusicoccin (FC), which can mimic the effects of auxin, can also stimulate germination, although auxin itself has no influence on germination. The lack of convincing evidence in support of a role for PGRs in germination may reflect the inadequacies of the experimental approaches that have been adopted. Currently, immunological tech-

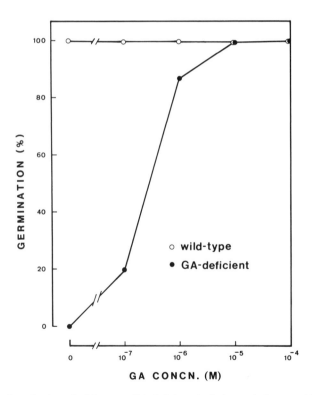

Figure 6.3 Germination of wild type or GA-deficient *Arabidopsis thaliana* seeds after exposure to a range of concentrations of GA_{4+7}. (Redrawn from Karssen *et al.*, 1987.)

niques for the detection of PGRs are emerging, and these should enable researchers to focus their efforts on a determination of the auxin, GA, CK, and ABA status of embryo cells and tissues at the time of germination. Such experimental approaches should prove illuminating.

6.4 Mobilization of storage reserves

Mature seed contains a wealth of stored food reserves laid down during development in order to sustain the young seedling during its early growth. Some of these reserves will also undoubtedly be required during prolonged periods of dormancy, when the imbibed seed synthesizes RNA, protein and membrane phospholipids. These reserves comprise carbohydrates, lipid, protein, phosphorus and mineral ions, and are located in the embryonic axis, but also and more predominantly in storage organs such as the endosperm and cotyledons. In the case of cereal grain, there is overwhelming evidence that PGRs regulate the mobilization of storage reserves. In other seeds such evidence is meagre. This is possibly due to the difficulty of isolating component tissues and organs of non-cereals for investigations, a process which is straightforward in cereal grain as a result of their spatial separation. Alternatively, it may be a feature restricted to seeds such as cereals which have a clear 'digestive tissue', the aleurone, which secretes enzymes into dead storage tissue (see Figure 6.4).

It has become widely accepted that during the early hours of germination in cereals, GAs produced by the embryo and/or its scutellum are released into the endosperm where they diffuse to the aleurone layer. Here they induce the production and secretion of a number of hydrolytic enzymes capable of breaking down the reserves in the endosperm (see Baulcombe *et al.*, 1986). The most abundant of these enzymes is α-amylase. Evidence for the involvement of GA in the control of these events is remarkably simple. Removal of the embryo dramatically slows down endosperm hydrolysis, while incubating isolated embryos and embryectomized seed together restores α-amylase levels. A low-molecular-weight diffusible factor has been isolated and appears to be a GA. The facts that GAs can induce endosperm degradation in embryoless grain and that transient increases in GA-like activity commonly occur in barley prior to α-amylase production, support the role proposed for the PGR. Nevertheless, it is now clear that additional signals other than GAs are also involved in controlling aleurone cells. Furthermore, in addition to producing and releasing GAs, cereal embryos also synthes-

Figure 6.4 Scanning electron micrograph (SEM) of longitudinal median section through an *Avena fatua* grain showing endosperm (*EN*), embryo (*E*) and aleurone (*AL*). Inset: higher-power SEM of aleurone. (Photographs courtesy of Dr John Sargent.)

ize and secrete α-amylase (see Akazawa and Miyata, 1982), and this may further explain why their removal retards endosperm hydrolysis. It is not yet known how great the contribution of α-amylase is from the embryo during germination, and this is currently a subject of debate.

Controversy has also surrounded the role of endogenous GAs in the induction of α-amylase. However, after the results of a recent study (Gilmour and MacMillan, 1984), the balance of opinion has settled on the proposal that GA_1 and perhaps GA_3 are released by germinating cereal embryos and act directly on the aleurone cells.

Isolated aleurone layers respond to GAs by secreting a range of hydrolytic enzymes; α-amylase is the most abundant and therefore most extensively studied of these. Aleurone layers isolated from 3-day imbibed half-seeds (grains with their embryos excised) of Himalaya barley begin to secrete α-amylase approximately 8 h after addition of GAs (Figure 6.5). Aleurone layers of wheat and oat have longer lag times. Following this is a phase of sustained synthesis and secretion of α-amylase lasting for several days. Density labelling experiments with

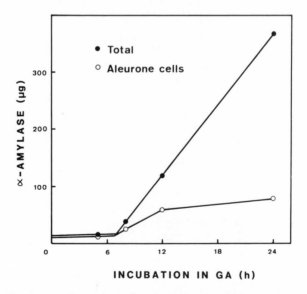

Figure 6.5 Time-course of α-amylase release by barley aleurone layers after incubation with $1\,\mu M$ GA. Enzyme activity was measured in the medium surrounding the aleurone layers or in the aleurone cells themselves. Total refers to the sum of both these activities. (Redrawn from Bewley and Black, 1982.)

$H_2^{18}O$ have proved that the enzyme is synthesized *de novo*. GA treatment of aleurone cells stimulates poly(A^+) polymerase activity and incorporation of labelled adenine into poly(A^+) mRNA. Concomitant with this stimulation of mRNA synthesis is a specific accumulation of α-amylase mRNA. 'Run on' transcription experiments performed with nuclei isolated from aleurone protoplasts indicate that the increase in α-amylase mRNA induced by GA is precipitated by an increase in the rate of transcription of α-amylase genes, and that ABA prevents this (Jacobsen and Beach, 1985).

GAs stimulate the activity of a number of other enzymes in aleurone cells. The production and secretion of proteases, for example, closely resembles that of α-amylase and it has been suggested that they may be co-secreted. Secreted proteases are probably responsible for hydrolysing endosperm protein reserves. A β-1,3-glucanase is synthesized by aleurone cells in the absence of GAs, and the PGR stimulates its secretion from aleurone cells. This enzyme is thought to degrade endosperm cell walls, rendering the starch and protein accessible to their respective hydrolases. A number of other enzymes fall into this category, in that GAs control their secretion but not their synthesis. These include ribonuclease, peroxidase and acid phosphatase.

In view of the dramatic effect of GAs in promoting secretion of endogenous and newly synthesized proteins from aleurone cells, it is not suprising that fine-structural studies show that the PGR brings about substantial changes in the endomembrane system. For instance GAs induce extensive fenestration and vesiculation of ER cisternae. This results in shorter profiles of ER in GA-treated aleurones with associated ER-derived vesicles. These changes are associated with elevated turnover of phophatidylcholine to triacylglycerols in the aleurone cells which may be mediated by increases in the activities of enzymes of the cytidine diphosphate-choline pathway. These biochemical observations support the hypothesis that GAs stimulate endomembrane flow within the aleurone cells. α-Amylase has been detected in endoplasmic reticulum and dictyosomes, but at present the precise secretory route is not known.

The aleurone layer is the major site within the cereal grain where phosphorus and the mineral ions calcium, magnesium and potassium are stored. They occur as phytin, the mixed Ca^{2+}, Mg^+, K^+ salt of myoinositol hexaphosphoric acid, which is present as a globoid within aleurone protein bodies. Although phytin hydrolysis occurs independently of GAs, the PGR is involved in regulating the release of the phosphorus and mineral ions from aleurone cells. Aleurone protein bodies are bounded by a single bilayer membrane through which these reserves must pass in order to enter the cytoplasm. It is possible that GAs influence the transport capacity of this membrane, and in accord with this hypothesis it has been reported that radiolabelled GAs will bind to aleurone protein bodies.

6.5 Conclusions

Our understanding of the role of PGRs during seed development, dormancy and germination is at best fragmentary but encouragingly is advancing at an increasing pace. Future progress will be fuelled by the opportunities afforded by the new experimental techniques of molecular biology and the study of developmental mutants. For instance, monoclonal antibodies raised against PGRs will enable us to characterize the true status of these compounds at the cellular level. However, we also need to strive towards being able to assess the responsiveness of plant tissues to their endogenous PGRs, and in this respect a greater understanding of PGR receptors and the events subsequent to receptor occupancy is required. The exciting progress that has been made with the cereal aleurone system illustrates how far our understanding of PGR action can develop, given an experimental system which is amenable for study. Furthermore, it undermines the argument that dismisses a strategic role for PGRs in plant development on the basis of data generated from intractable systems. The problems of seed development, dormancy and germination have so far proved more recalcitrant and have yielded few clues about PGR action. They should therefore be viewed not as areas to be avoided but as areas of challenge for future research strategies.

CHAPTER SEVEN

ROOT AND SHOOT DEVELOPMENT

After germination, the further development which the root and shoot undergo ultimately dictates the final form of the mature plant. The nature and extent of these developmental changes vary considerably between species, while individuals within a single species tend to conform to certain guidelines. The extent of these restrictions are the culmination of an interaction between a plant's genotype and its environment. In this and the next chapter we examine the evidence that PGRs play a role in influencing the phenotypic expression of a plant, and investigate the possible mechanisms by which this is achieved.

7.1 Growth

Plant growth is a function of cell division and cell expansion. The process of division is limited to specific meristematic regions, whereas expansion occurs in cells throughout the body of the plant and is the primary means by which a plant increases in size. The rate of growth is largely dictated by environmental factors such as light, temperature, nutrients and water availability; however, it can be influenced by the application of PGRs. This information is significant not only because it implicates PGRs in the regulation of growth *in vivo*, but also because an ability to manipulate growth is of major interest to the horticultural and agricultural industries (see Chapter 11).

The driving force behind cell expansion is the uptake of water by osmosis. This process is restricted by such factors as cell wall extensibility, the concentration of cytoplasmic solutes, and the permeability of the cell to water. The relationship between growth and these three factors can be summarized by the following equation (Cosgrove, 1986):

$$V_s = \frac{L\phi}{L + \phi} (\gamma\Delta\pi - Y)$$

Thus at a 'steady state' of growth, the rate of cell expansion V_s is a function of the hydraulic conductance of the cell L; the wall extensibility ϕ; the Reflection Coefficient (γ), which is a measure of the selectivity of the plasmamembrane towards a particular solute; the difference in osmotic potential across the plasmamembrane ($\Delta\pi$); and the yield threshold of the cell wall (Y), which is the minimum turgor required for wall expansion. Formulation of expansion growth into an equation of this kind helps to focus on those parameters that may have a key role in the regulation of growth and which may therefore be targets for the action of a PGR.

The contribution made by hydraulic conductance to growth regulation is the subject of controversy. In pea segments, it is the magnitude of wall extensibility and not hydraulic conductance which restricts growth. However, the results from studies on soybean hypocotyl are not as conclusive. Furthermore, the demonstration that IAA can enhance the rates of plasmolysis and deplasmolysis of onion epidermal cells implies that PGRs could influence growth via hydraulic conductance. Irrespective of this, there is overwhelming evidence that the primary effect of PGRs is on the extensibility of cell walls. This biophysical parameter can be readily quantified in tissue segments using an Instron tensiometer. From the stress–relaxation curves obtained, the elastic (reversible) and plastic (irreversible) extensibility can be determined. Both IAA and GAs promote the plastic extensibility of stem tissue, whereas only IAA appears to enhance the elastic component. Ethylene has a dual effect on wall extensibility, inhibiting it in the longitudinal direction but increasing it laterally. These results are in accord with the capacity of ethylene to inhibit elongation and promote lateral expansion growth in tissues such as pea epicotyl.

Information on the nature of the mechanism governing wall extensibility can be obtained by studying the kinetics of elongation growth after PGR treatment. Typically, the growth of coleoptile or stem segments is stimulated within 10–20 minutes of IAA or GA application. The rate of IAA-promoted growth is commonly maximal after 30 minutes and then declines, although subsequently a further increase may occur (Figure 7.1). It has been suggested that this deceleration nestled between the two peaks of growth may represent a stage where osmoregulation lags behind cell expansion (Vanderhoef and Dute, 1981). Similarly rapid kinetics are exhibited by the ethylene-inhibited growth of pea epicotyl and radish root tissue (Eisinger, 1983). The speed of the growth response to auxin prompted the suggestion that elongation could not be mediated by differential gene expression; however, this supposition has now been proved incorrect by the demonstration of Theologis *et al.* (1985) that IAA can induce the accumulation of specific mRNAs in pea

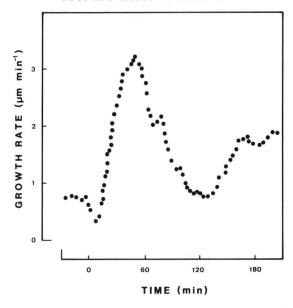

Figure 7.1 Time-course of growth rate of *Vigna radiata* hypocotyl segments in the presence of IAA (10^{-4}M). (Redrawn from Bouchet *et al.*, 1983.)

epicotyl tissue prior to the onset of elongation. The identity of the proteins encoded by these mRNAs remains unknown.

Although IAA promotes wall extensibility, the initial site of action of the PGR is not the wall itself but the intact protoplast. It has therefore been suggested that a 'wall loosening factor' must move from the cell into the wall where it acts to promote extensibility. This line of thought led to the proposal of the 'acid growth theory' of cell elongation in 1971. According to this hypothesis, auxin stimulates plant cells to excrete protons into the cell wall where they promote wall loosening. The beauty of this simple idea is that it makes specific predictions that can be readily tested. Clearly IAA must stimulate proton efflux, but also low pH should promote wall loosening, neutral buffers should inhibit IAA-induced growth, and other compounds that enhance proton excretion such as the fungal toxin Fusicoccin (FC) should elevate elongation. Apart from problems encountered in the detection of proton efflux in some tissues due to insulation by the cuticle, the balance of published results have come out in favour of the theory. However, a recent study on maize coleoptiles by Kutschera and Schopfer (1985) has re-examined the predictions made by the 'acid growth theory' strictly on a quantitative rather than qualitative basis. The conclusions from their results are that FC does fulfil the predictions made by the theory but IAA does not. Much of the conflict in this area may originate from the preoccupation of studying populations of cells which exhibit a heterogeneous response to IAA. Until a model system of homogeneous cells or protoplasts is examined, it is unlikely that the present confusion will be fully resolved.

Although there are abundant reports linking proton extrusion with IAA action, demonstrations of an effect of GAs on wall pH are rare. Taiz (1984) has suggested that GAs may act by promoting Ca^{2+} influx, and that the ratio of H^+ to Ca^{2+} in the extracellular space rather than the level of protons *per se* may govern wall extensibility. If this were correct it would provide an explanation for the ability of high Ca^{2+} levels to inhibit both GA- and IAA-stimulated elongation and for the change in capacity of the cell wall to undergo acid-induced loosening during development. Further support for a role for Ca^{2+} in expansion growth is the recent demonstration by Cleland (1986) that Ca^{2+} inhibits, and the Ca^{2+} chelator EGTA promotes, wall extensibility of soybean hypocotyl sections (Figure 7.2).

Ethylene-induced lateral expansion is prevented by neutral buffers and vanadate (an inhibitor of proton efflux). This implies that ethylene-promoted growth is mediated by a similar mechanism to that of IAA. Moreover, growth of the water plant *Nymphoides* is stimulated by both ethylene and IAA in an additive fashion, and this is mimicked by their combined capacity to promote proton efflux. Since a root has no cuticle, subtle changes in acidity or alkalinity can be monitored by arranging beads containing pH indicators along its length. This technique has been used successfully to demonstrate that elongation growth in roots is also associated with a reduction in wall pH, which can be stimulated both by the application of FC and by gravistimulation. The method can also be used to apply PGRs at specific locations along a root so that localized growth changes may be monitored.

Proton efflux induced by IAA treatment is suppressed by cycloheximide, cerulenin, or vanadate; inhibitors of translation, fatty acid biosynthesis and ATPases respectively. Demonstrations of a direct stimulation of ATPase activity *in vitro* by IAA have been infrequent. This may be because IAA acts by decreasing the apparent K_m of the enzyme, and therefore at high ATP concentrations no effect of auxin would be apparent (see Chapter 10). The cellular location of the ATPase that is stimulated by IAA is unknown. Vanadate is thought to specifically

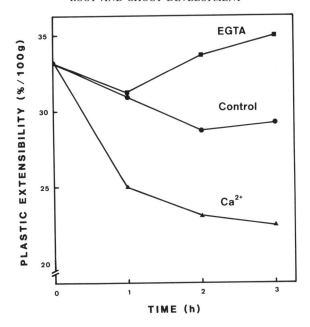

Figure 7.2 Effect of the addition (Ca^{2+}) or removal (EGTA) of calcium ions on the wall extensibility of soybean hypocotyl cell walls. (Redrawn from Cleland, 1986.)

block the plasmamembrane bound enzyme, although the requirement for fatty acid biosynthesis to maintain auxin action implies that the ATPase might be associated with a cytoplasmic vesicle which requires frequent replenishment. In accord with this proposal, Theologis has recently (1986) put forward a scheme to account for IAA-stimulated proton efflux, based on the premise that the ATPase is associated with the Golgi structure. Furthermore, he proposes that proton pumping relies on the translation products of some of the early mRNAs induced by IAA (Figure 7.3).

Whilst the identity of the signal which passes from the protoplast to the cell wall (or perhaps vice versa) remains speculative, it is widely accepted that the changes in wall extensibility induced by PGRs are the culmination of enzyme action. In the absence of a precise model of cell wall structure, one can only make an educated guess as to the nature of the stress-bearing bonds responsible for extensibility and the enzymes that might be involved in their cleavage. Fry (1986) has proposed that the cell wall structure is maintained by covalent cross-links in

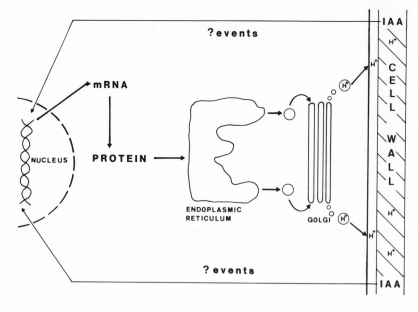

Figure 7.3 Theologis' mechanism of IAA-induced H$^+$ efflux based on the hypothesis that induction of mRNA synthesis is the primary effect of the auxin. It is suggested that IAA induces the synthesis of a protein which facilitates the fusion of secretory vesicles containing cell wall materials with the plasma membrane. The stimulation of H$^+$ efflux by the auxin is proposed to be associated with the secretory vesicles rather than playing a direct role in the growth-regulating process. (After Theologis, 1986.)

the form of glycosidic bonds, diferuloyl bridges or interpolypeptide linkages formed by isodityrosine. Since peroxidases may be involved in catalysing the cross-linking of both ferulate and tyrosine, a suppression of the activity of this enzyme would reduce the rate of rigidification of the cell wall. Interestingly, both IAA and GAs suppress peroxidase activity in pea epicotyl tissue, whereas ethylene promotes the activity of the enzyme. However, a change in rigidification would not alter the plastic extensibility of the cell wall, although it might affect the elastic component. The activity of the glycosidic enzyme β-galactosidase in *Avena* coleoptiles is stimulated by IAA and inhibited by neutral buffers, which makes it a promising candidate to play a role in the regulation of wall extensibility. Unfortunately, the enzyme does not correlate closely with growth in other tissues, nor is elongation suppressed if β-galactosidase activity is suppressed by aldonolactone. Dextranase activity is also promoted in *Avena* coleoptiles by auxin, although the kinetics of its rise in activity appear to be too slow to account for the rapid response to IAA. However, since the enzyme is a transglycolase and thus has the capacity to cleave glycosidic links and resynthesize them at new locations, it is possible that this enzyme is involved in the long-term growth responses to IAA. Other enzymes that might be involved in growth include those for wall biosynthesis, since the thickness of the cell wall does not decrease during elongation. In pea stem tissue, both IAA and FC stimulate the activity of UDPG-β-1,4-glucan synthase, an enzyme which regulates polymer biosynthesis. Furthermore, the activity of UDPG-β-1,4-glucan synthase is stimulated within 20 minutes of exposure of the tissue to IAA and this increase is suppressed by vanadate.

Much of the work described has been carried out on tissue segments. This material is ideal to work with since PGRs can be readily applied and the plant tissue responds rapidly due to a deceleration in the growth of 'untreated' segments as a result of excision. Although GAs have the capacity to stimulate the growth of intact plants, the effect of IAA on whole plants is minimal. This has prompted Hanson and Trewavas (1982) to argue that the growth-promoting effects of IAA on tissue segments largely represent an acceleration of the recovery from the injury caused by excision. Although this might represent part of the story, the growth of intact pea plants is stimulated at a comparable rate to that of segments during the first 6 h after IAA treatment. By 24 h, control plants 'catch up', which may account for the discrepancies in reports of the effects of auxins on intact plants.

In this section we have described the effects of PGRs on growth and discussed the mechanisms by which cell expansion may be regulated. The most crucial question remains; do PGRs regulate growth *in vivo*? It is appropriate at this stage to consider what might be the most convincing approaches by which to address a question such as this. One way to ascertain the role of a PGR is to manipulate the endogenous level of the compound and examine the effect this has on the phenotype of the plant. This might be achieved by the application of chemicals which specifically affect PGR biosynthesis, or more elegantly, by the characterization of PGR mutants. Ideally, it should be possible to supplement a PGR-deficient plant, obtained either genetically or chemically, with applied PGR so that it reverts to the 'normal' phenotype. Application of a variety of inhibitors of GA biosynthesis to plants has been reported to reduce growth, and a number of these compounds have been exploited as commercial growth retardants (see Chapter 11). However, as yet no specific inhibitors of IAA biosynthesis have been identified. A number of dwarf mutants of maize, pea and tomato have been characterized which revert on the application of exogenous GAs (Stoddart, 1987). These dwarfs have been found to have abnormally low levels of GA_1 as a result of blocks at specific sites in the biosynthetic pathway of this PGR (see Figure 7.4). This is convincing evidence that GA_1 plays a critical role in stem elongation *in vivo*. Other dwarf mutants have been classified as 'GA insensitive', since they cannot be reverted by GA applications and their endogenous GA_1 level is normal. It remains to be seen whether such plants prove to be GA-response mutants or to have their growth restricted in other ways. Paradoxically, no dwarfs have been characterized as being IAA-deficient. This may imply either that such mutations are lethal, or that IAA does not directly regulate stem growth *in vivo*.

7.2 Tropisms

If the rate of growth on either side of a root or shoot is unequal, then the organ bends. Such tropic responses can be induced by a range of environmental stimuli including gravity, unilateral light, and mechanical perturbation, and PGRs have been implicated in major roles in the regulation of each of these.

7.2.1 Gravitropism

It is widely accepted that gravity is perceived by amyloplasts which act as statoliths by sedimenting to the bottom of specific statocyte cells. In roots, these cells are restricted to the columella region of the root cap, while in shoots they are distributed throughout the length of the organ and occupy a position in the endodermis adjacent to the vascular tissue. On gravistimulation, amyloplast movement occurs almost immediately, with bending taking place after a lag period of about 10 minutes. Bending is commonly achieved by a diminution or cessation of growth on the concave surface of the root or shoot, while only occasionally does the growth rate on the convex surface increase. Consideration of this information is important in the development of any theory to account for the mechanism of tropic curvature.

The most durable theory of gravitropism was proposed independently by Cholodny and Went in 1926. These workers hypothesized that auxin moved from the apex into the elongating zone, where it became asymmetrically distributed as a result of lateral transport towards the lower side of the root or shoot. The elevated level of auxin on the lower side

Figure 7.4 The effect of GA_1 on the growth of the gibberellin-deficient maize mutant *dwarf*-2. The *dwarf*-2 mutation appears to block the oxidation steps GA_{53}-aldehyde → GA_{53} and/or GA_{12}-aldehyde → GA_{12}. For further details see Phinney (1984). (Photograph reproduced from Phinney and Spray, 1982, with permission.)

was optimal for shoot growth and therefore promoted elongation, but supraoptimal for root growth which was thus inhibited. The Cholodny–Went hypothesis has received considerable support from studies on shoots and coleoptiles. With the aid of radiolabelled IAA it has been possible to follow the progress of auxin movement during the gravitropic bending of both excised and intact tissues. Gravistimulation results in the asymmetric distribution of applied ^{14}C-IAA in a range of shoot tissues including *Zea mays* coleoptiles and *Helianthus annuus* hypocotyls, with a greater proportion of the label migrating to the lower side. Moreover, an increase in the percentage of endogenous IAA has been found in the lower half of maize mesocotyl tissue within 15 minutes of gravistimulation (Bandurski *et al.*, 1984). A similar distribution of IAA has been identified in gravistimulated coleoptiles using a sensitive radioimmunoassay procedure, although this technique was unable to detect an asymmetry in auxin levels in *Helianthus* tissue (Mertens and Weiler, 1983).

The majority of reports of stem tissue asymmetries indicate that after gravistimulation the distribution of auxin is about 2:1 in favour of the lower side. Objectors to the Cholodny–Went theory argue ardently that the magnitude of this difference is not sufficient to account for the observed rates of bending, while 'defenders of the faith' seek sanctuary in changes in subcellular compartmentation as a plausible explanation for this apparent shortfall in the theory. Until the localization of the changes in IAA levels are pinpointed, this controversy will continue to rage. Quantification of the IAA within the epidermal cells may prove to be particularly decisive, since it is likely that the rate of elongation of these cells regulates bending. Paradoxically, although the evidence that GA plays a significant role in the regulation of straight growth is convincing, reports of a redistribution of this PGR during gravitropic bending are rare. Furthermore, there is no evidence that GA-deficient mutants respond abnormally to gravity. If gravitropic bending is primarily regulated by the epidermal tissue, it is conceivable that the elongation of these cells *in vivo* could be under the control of IAA and not GA.

It is evident that IAA can become redistributed in stems or coleoptiles irrespective of whether it is the cause or effect of gravitropism. The discovery that binding sites for auxin efflux may be localized within the endodermal cells (see Chapter 9) has raised the possibility that sedimentation of amyloplasts may have a direct effect on IAA transport. In support of this hypothesis is the demonstration that the polarity of IAA transport in nodes of *Echinochloa colonum* can be reversed by gravity (Wright, 1981). Bandurski *et al.* (1986) have proposed that amyloplast sedimentation deforms the bioelectric field of the epidermal cells and causes plasmodesmata on the lower side to open, allowing IAA to move from the stele into the epidermal cells via the cortex. Since the intercellular channels on the upper side would remain closed, no auxin would flow into the upper cortical and epidermal cells, and growth in these tissues would rapidly diminish or cease altogether. Whilst the basic tenets of this theory have yet to be critically appraised, the predictions that it makes concerning growth rate changes during gravitropism are in agreement with those that are observed.

Although the majority of effort has been centred on studies of the distribution of PGRs during gravitropism, there have been a number of reports that ions such as Ca^{2+} may become redistributed in stems in favour of the upper side. In view of the ability of Ca^{2+} to inhibit elongation growth (see section 7.1) it is possible that the ion plays a significant role in gravitropism. Alternatively, its redistribution maybe a consequence of the 'geoelectric effect' which may be caused by the asymmetrical distribution of IAA.

The gravitropic mechanism in grass nodes may be different from that in shoots, since the leaf sheath base is 'quiescent' prior to gravistimulation. Growth can only be induced in the tissue by reorientation or treatment with IAA. Since excised nodal half-segments from *Echinochloa colonum* can be induced to grow when placed horizontally with the epidermis lowermost, it is unlikely that the gravitropic response relies on the lateral redistribution of a PGR. However, within 30 minutes of placing in this orientation, the IAA content of the tissue increases by 40%, and within 1 h the ethylene production by the nodal tissue is also enhanced. The most likely explanation for the latter observation is that the rise in endogenous IAA stimulates the biosynthetic pathway of ethylene (see Chapter 4). The gas itself appears to play no role in the gravitropic response, since its production can be inhibited by AVG without altering the rate of bending. If sedimenting amyloplasts can induce bending in nodal tissue through a direct stimulation of IAA levels, then perhaps they have the capacity to act in a similar way in other tissues.

As it stands, the Cholodny–Went hypothesis is difficult to reconcile with the acropetal polarity of IAA transport found in roots. An alternative explanation might be that gravity stimulates the redistribution of a growth inhibitor, the source of which lies in the root cap. This hypothesis is supported by the demonstration that removal of the cap

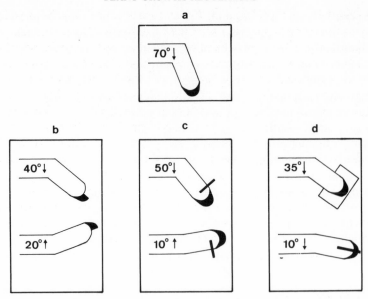

Figure 7.5 Curvature of *Zea mays* roots exhibited 12h after being placed in a horizontal position (arrow indicates direction of bending). (*a*) Intact root; (*b*) half root cap removed; (*c*) impermeable barrier inserted unilaterally behind the apex; (*d*) barrier inserted longitudinally in the apex. These results support the hypothesis that the root cap contains a growth inhibitor and that this compound is translocated towards the lower side of the root tip after gravistimulation. (Redrawn from Shaw and Wilkins, 1973.)

stimulates root growth. Furthermore, in a classic series of surgical manipulations on *Zea mays* roots, Shaw and Wilkins (1973) have demonstrated that a putative inhibitor is translocated in the cap to the lower side upon gravistimulation (see Figure 7.5). Although asymmetrical application of ABA will induce bending, it is unlikely that ABA is the inhibitor, since it is not consistently redistributed during bending, and chemical or genetic means of suppressing ABA biosynthesis do not affect the gravitropic response. The identity of the inhibitor may well emerge from studies on mutants of maize whose roots require a light stimulus before they will bend. Light is perceived by the cap, and excised caps can be irradiated and will induce a graviresponse when placed horizontally on detipped dark-grown roots. A number of changes have been shown to take place in the excised cap tissue on illumination, one of which is the appearance of a neutral growth inhibitor. The identity of this compound is unknown.

Recently a model to account for root gravitropism has been put forward, centred on a key role for Ca^{2+}. The basis of this theory is that chelation of Ca^{2+} inhibits bending, and that Ca^{2+} is redistributed to the lower side of the root cap within 30 minutes of gravistimulation. The proponents of the theory suggest that Ca^{2+} sensitizes the tissues to the growth-suppressing action of IAA. This hypothesis highlights one of the problems that must be contended with when proposing a mechanism for gravitropism in both roots and shoots. That is, that as a result of placing a tissue horizontal a variety of ions and molecules may be redistributed. Therefore, we must separate those changes that are causally related to gravitropism from those that are consequently or casually related. Moreover, the link between graviperception and the redistribution of ions or molecules must be established. It is possible that progress on both of these fronts will be achieved by a comparative study of normal plants and some of the gravitropic mutants which have recently been characterized (Roberts, 1987).

7.2.2 Phototropism

Phototropic bending is induced by illuminating plant material with unilateral light. In general, shoots grow towards a light source while roots grow away from one. The majority of the research that has been carried out has centred on the first positive phototropic response of etiolated coleoptiles which occurs at a fluence of less than $10^{-1} J m^{-2}$. Our knowledge of the mechanism of phototropic bending of green shoots growing in the natural environment is meagre, and of roots almost non-existent.

In coleoptiles the site of maximum sensitivity to unilateral irradiation is located just below the tip. Phototropic responses induced by illuminating tissues below the apex are not the result of light piping, since unilateral irradiation stimulates significant bending even after the apical 5mm of the coleoptile has been excised. In green plants it is important to separate those tissues that contribute to the growth response from those that might house the phototropic receptor. On balance, the evidence favours the apex and leaves in the former category, and hypocotyl or epicotyl tissue in the latter. Unlike gravitropism, no single pattern of growth-rate changes appears to predominate in causing bending; furthermore, the growth response of individual cells is dependent on the existence of a light gradient across the organ rather than the absolute quantity of irradiation on either side of it.

The first demonstration that unilateral illumination of coleoptiles resulted in an asymmetric distribution of auxin was made by Went in 1928. The technique that he employed was to collect diffusates from phototropically stimulated coleoptile tips into agar blocks and measure auxin activity in the blocks by bioassay. Modern approaches utilizing radiolabelled IAA have upheld Went's original observations, and have indicated that approximately twice as much auxin accumulates on the shaded compared to the illuminated side of a coleoptile. The asymmetry of auxin levels induced by phototropism is similar to that caused by gravity, and comparable reservations have been expressed about the inability of these values to account for the differential growth observed. The phototropically-induced asymmetry in auxin can be prevented if the coleoptile tips are bisected longitudinally by an impermeable barrier. This result implies that the asymmetry in auxin distribution is the result of a lateral translocation of the PGR towards the shaded side of the coleoptile, rather than a reduction in auxin biosynthesis or a stimulation of auxin destruction on the illuminated side. Unequivocal evidence in support of this proposal has come from studies of the distribution of ^{14}C-IAA applied asymmetrically to the extreme apex of phototropically stimulated maize coleoptiles. Reports of an asymmetry in the distribution of other PGRs after phototropic stimulation are rare. In light-grown sunflower hypocotyl tissue, no redistribution of auxin has been detected after unilateral illumination; however, a twofold increase in the level of the growth inhibitor xanthoxin acid has been recorded on the illuminated side. Interestingly, phototropically stimulated tips of the same plant have been reported to exhibit a 10-fold increase in the level of diffusible GA-like activity on the shaded side. In view of the overwhelming evidence that GAs play an important role in plant growth, this observation should perhaps be followed up.

The controversy as to whether the asymmetrical distribution of auxin is the cause or result of bending is as great in phototropism as it is in gravitropism. Whether the differential growth induced by these two environmental stimuli turns out to be mediated via PGRs, ions, or even electrical fields, it is tempting to speculate that the mechanism responsible for both will turn out to be the same, with only the transduction of the initial stimulus being different.

7.3 Apical dominance

The apical tissues of a root or shoot have a profound influence on the form that a plant adopts. These regions have the ability to restrict the development of lateral bud meristems and maintain them in a quiescent state until the root or shoot apex loses its potency to dominate, or is removed from the plant by accident or design. The capacity of a plant to sustain a supply of replacement apices is not only of major ecological advantage, but is also an excellent example of the way in which plant growth and development are integrated. In addition, the phenomenon is of significant commercial interest to the agricultural and horticultural industries, since the axillary outgrowth may have a direct influence on such factors as the ability of a plant to intercept solar radiation and take up nutrients and water, and its accessibility for applications of herbicides and pesticides.

Since decapitation of a root or shoot stimulates the outgrowth of lateral meristems, it is evident that the apex is the source of some correlative signal. Progress on its identification from shoot apices was first made by Thimann and Skoog (1933). These workers allowed the contents of excised *Vicia faba* apices to diffuse into agar blocks, and when these were replaced on decapitated *Vicia* plants, dominance was reimposed. The active constituent in the diffusate was later identified as auxin, and the ability of IAA to inhibit lateral outgrowth has been subsequently confirmed for a range of plants. To complete the picture, IAA has been identified in the apex and young leaves of a number of species. If IAA is the correlative signal, then the PGR could act in one of two ways. Firstly, it could move from the apex to the lateral meristem where it could have a direct inhibitory influence on growth. This hypothesis is supported by the demonstration that the auxin transport inhibitor TIBA promotes the outgrowth of lateral buds situated basipetal to its site of application. However, the compound also induces abscission of the apical leaves and thus may overcome dominance by chemically decapitating the plant. Moreover, measurements of the movement of ^{3}H-IAA or the radiolabelled synthetic auxin ^{14}C-2,4-dichlorophenoxyacetic acid (2,4D) applied to the decapitated stumps of *Phaseolus vulgaris* or *Helianthus annuus* plants have failed to detect label in the developing lateral buds, and therefore a direct effect of auxin seems unlikely. The alternative is that auxin inhibits lateral bud growth indirectly. Compelling evidence in favour of this has come from studies by Brown et al. (1979) on sunflower (Figure 7.6). These workers observed that within 1 h of applying ^{14}C-2,4D to the stump of a decapitated seedling, label was distributed throughout the plant, but by 24 h the majority of the label had migrated back to the stump apex. Excision of the stump tissues at this time released the laterals from apical dominance. If insufficient

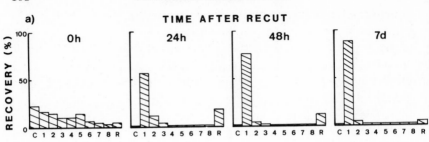

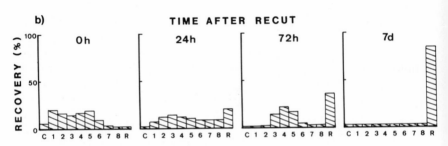

Figure 7.6 The distribution of ^{14}C-2,4-D in decapitated *Helianthus annuus* seedlings following an application of the synthetic auxin which (*a*) was sufficient to maintain dominance, or (*b*) was insufficient to maintain dominance. In both experiments the apical 15mm of stem was recut and discarded 1h after 2,4-D application to give a $t = 0$h value. Recovery of radiolabel was monitored in 10mm segments, segment 1 being closest to the shoot apex. *C*, cotyledon; *R*, root. (Data replotted from Brown *et al.*, 1979.)

2,4D was applied to decapitated seedlings to maintain dominance, then by 24 h the majority of the label had accumulated in the roots. In addition, there was a correlation between the ability of 2,4D to inhibit lateral outgrowth and its ability to stimulate callus growth at the stump surface. This series of observations imply that as long as a sufficient concentration of auxin is applied to stimulate the maintenance of a sink at the apex, then this PGR will inhibit lateral outgrowth and direct transport. If the concentration of auxin is not great enough, after decapitation, the root apices become the major sink and the PGR is directed towards them.

Auxin is not the only PGR which can influence the outgrowth of laterals from plant shoots. Application of cytokinins to quiescent lateral buds from a range of species stimulates their growth. Commonly the effects of cytokinins are transitory, lasting for only a few days, but outgrowth can be prolonged by combined treatments with IAA. It is difficult to estimate the role of cytokinins *in vivo* since no detailed analysis of the levels of this PGR in lateral buds during the release from dominance has been undertaken. The possibility of a role for ABA in the inhibition of lateral growth has been explored, since ABA can suppress lateral development. Although a few correlations based on bioassay data have appeared in the literature, these have been challenged after re-analysis of ABA levels using HPLC techniques; furthermore, ABA-deficient mutants have not been reported to exhibit weakened dominance. It is unlikely, therefore, that ABA makes a major contribution to the maintenance of apical dominance in shoots. In contrast, there is a considerable body of circumstantial evidence that ethylene might play a role in this phenomenon. For instance, exposure of apical tissues to ethylene or constriction of the apex (which elevates ethylene production), releases laterals from inhibition. Additionally, lateral outgrowth in decapitated plants is inhibited by AVG or Ag^+, although localized application of ethylene to laterals from intact plants does not release them from dominance. Although such results implicate ethylene in apical dominance, the ability of this PGR to inhibit auxin transport (see Chapter 4) may contribute to its effects. Certainly the gaseous nature of ethylene makes it an unlikely candidate as a correlative signal.

In comparison to shoots, the dominance in root systems is a relatively unexplored phenomenon. Surgical manipulations of pea seedlings indicate that the root tip is the source of a lateral root inhibitor and the cotyledons of a promoter. Although the identity of these compounds is unknown, it is plausible that they are PGRs, since application of auxin to pea roots promotes lateral root initiation (possibly via ethylene), while CKs and ABA inhibit the process. However, until a detailed analysis of PGRs is carried out, the evidence remains unconvincing.

Our current state of knowledge of apical dominance favours a key role for auxin in shoots and possibly other PGRs in roots. Whether dominance in these systems is imposed in the same way is unknown, but the concept that auxin acts at the shoot apex by diverting molecules towards it is appealing. These molecules may include nutrients and PGRs, and the directed movement of such compounds to the apex would effectively starve the laterals of the constituents necessary for growth. This 'nutrient diversion' hypothesis was essentially proposed by Went as long ago as 1938, and now with the aid of contemporary techniques of PGR analysis and molecular biology it is perhaps time for it to be re-investigated.

7.4 Bud dormancy and tuberization

Bud quiescence is not restricted to laterals, and in temperate herbaceous and woody perennials the shoot meristematic tissues are maintained in a dormant state throughout the winter months. The primary environmental factor which regulates the onset of bud dormancy is the length of the photoperiod. Lengthening days promote bud growth, whereas a decrease in the number of hours of light below a critical threshold induces the sequence of physiological and biochemical events leading to dormancy. Other environmental variables may influence the value of this threshold.

Working on the assumption that short days induced the production of a growth inhibitor in woody plants, Eagles and Wareing (1963) isolated substances from birch seedlings kept under short days which could suppress the growth of seedlings maintained under long days. The active principle was called 'dormin', and the major constituent was later identified as ABA. Paradoxically, although ABA was largely isolated through its association with bud dormancy, attempts to correlate its levels with this phenomenon in a range of species have been unsuccessful. It has been suggested that the amount of ABA synthesized in response to water stress can be influenced by photoperiod, and this may confuse attempts to correlate levels of this PGR with dormancy. Even if it turns out that ABA is not a critical regulator of bud dormancy, there is overwhelming evidence that it specifically induces the formation of resting structures or turions in the aquatic plant *Spirodella polyrrhiza*. Turion formation is induced over a narrow concentration range of ABA, and the PGR is effective only during a brief window in the development of the organism (Figure 7.7). Induction of turions by ABA is associated with the specific repression of nucleic acid and protein synthesis.

Involvement of gibberellins in the phenomenon of bud dormancy is well established. A decline in GA-like substances has been recorded in a range of species including birch and sycamore following their transfer to short days. In addition, treatment of some woody angiosperms with GAs can inhibit the dormancy-inducing capacity of short days. Few detailed analyses have been carried out on the effect of short days on GA levels in bud tissues, but photoperiod has been demonstrated to affect both the biosynthesis and the metabolism of GAs in spinach and pea tissues (Pharis and King, 1985).

Dormancy is broken after the bud tissues have experienced a period of low temperatures. Premature growth can be induced by the applica-

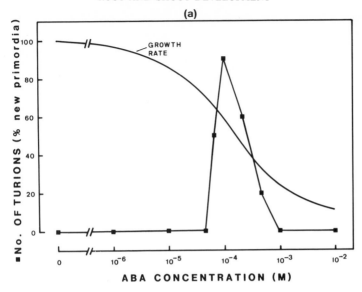

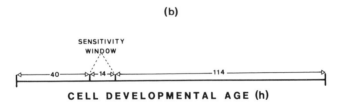

Figure 7.7 (*a*) The effect of ABA on growth and turion production in *Spirodela polyrrhiza*. (*b*) The period of maximum sensitivity to ABA for the turion forming response of *Spirodela polyrrhiza*. (Redrawn from Smart and Trewavas, 1983.)

tion of CKs to some woody plants and to dormant turions. Furthermore, there are a number of reports linking a rise in CKs with release from dormancy, and a decrease in the level of this PGR under shortening days. Thus we have at least three PGRs which potentially might regulate bud dormancy between them. However, in the absence of clear information about the molecular events which contribute to the making and breaking of bud dormancy, it is impossible to elucidate which (if any) PGR(s) are crucially important. The work that has been carried out on

Spirodella has demonstrated that the techniques are available to identify the biochemical changes that precipitate turion formation, and this approach has allowed the role of ABA in the induction of dormancy to be appraised critically in this plant. The time is now right to probe the induction of dormancy in higher plants in a similar manner.

Perpetuation of some plants over the winter period is achieved through the development of a dormant organ such as a tuber. Tuberization precedes the senescence of the remainder of the plant. In potato (and many other species), tuber formation, like bud dormancy, is regulated by daylength. The photoperiodic mechanism has some similarities to that which induces flowering in that there is evidence that both graft-transmissible inhibitors and promoters are involved (see Chapter 5). Grafting experiments carried out on potato indicate that a tuber-inducing substance originates from within the leaf tissues. Tuberization can be promoted *in vitro* by CKs; moreover, an increase in the level of *cis*-zeatin riboside has been reported to occur in stolons within 4 days of transfer of plants from long days to short days. No information is available as to whether this increase in CK activity is first detectable in the leaves. After development, potato tubers remain dormant for some weeks, and the duration of this quiescent phase is dependent upon both the cultivar under examination, and the temperature of storage. Dormancy can be broken by CK application, and there is convincing evidence that both the levels of this PGR and the sensitivity of the tuber tissue to it may regulate the duration of dormancy *in vivo* (see Chapter 4).

7.5 Conclusions

This chapter has presented an overview of some of the developmental processes which roots and shoots undergo during the life cycle of a plant. No attempt has been made to produce a comprehensive account of all the developmental events which take place in root and shoot tissues; rather, the role of PGRs in representative processes has been appraised. Although PGRs are implicated in all the events described, only the growth-regulating properties of GAs can be considered at this stage to be convincing. Ironically, it has become tacitly assumed that IAA is an endogenous regulator of growth, yet the evidence in support of this remains circumstantial. Moreover, in spite of the fact that there is a large body of evidence to support a role for PGRs in the regulation of tropic bending, apical dominance, and dormancy of root and shoot tissues, their involvement *in vivo* remains to be proved. At the end of the following chapter, some of the approaches that could be used to probe the role of PGRs in root and shoot development more critically are outlined.

CHAPTER EIGHT

LEAF, FLOWER AND FRUIT DEVELOPMENT

The aim of the previous chapter was to appraise critically the role of PGRs in root and shoot development. After a careful examination of a range of developmental phenomena, it is clear that there is sufficient evidence, albeit primarily circumstantial, to implicate PGRs in the regulation of growth, tropic bending, apical dominance and dormancy of root and shoot tissues. We now turn our attention to an assessment of the contribution that PGRs make during the growth and development of leaves, flowers and fruit.

8.1 Growth

The photosynthetic capacity of a leaf is proportional to the area of the lamina. Since photosynthetic capacity has a profound impact on the yield of a plant, there is considerable interest in studies on the regulation of leaf growth. Growth is a function of both cell division and enlargement; however, these processes can be temporally separated in a leaf, and therefore it is possible to probe their regulation individually.

Dark-grown leaves expand when illuminated. The light-induced leaf cell enlargement is accompanied by an increase in wall plasticity. In *Phaseolus vulgaris* leaves, growth commences within about 20 minutes of illumination, and is preceded by a decline in the pH of the leaf surface. If the leaf tissue is infiltrated with neutral buffers, growth is inhibited. These observations, coupled with the demonstration that acid treatment can promote leaf growth, imply that light-induced cell expansion in this organ is mediated by proton excretion. Thus the mechanism regulating growth in roots, shoots and leaves may be the same. As leaves age, their capacity to grow in response to a light stimulus declines. In bean leaves, this is the result of a reduction in the ability of the cell walls to undergo acid-induced loosening.

Although there are many reports that rapidly expanding leaves exhibit high endogenous levels of the shoot growth promoters IAA and GAs, this does not prove that these PGRs are involved in the regulation of leaf growth *in vivo*. In fact, treatment of leaf tissues with auxin has little effect on their growth unless the PGR is applied at high concentrations when malformations may be induced. In contrast, GAs can stimulate cell expansion, in bean leaf strips from plants grown under red light, within 30 minutes. GA-induced leaf growth is associated with an increase in cell wall loosening, but it is not inhibited by neutral buffers and is therefore unlikely to be mediated by proton extrusion. No information is available as to whether the effects of GAs are associated with a stimulation in Ca^{2+} efflux.

Removal of the root system of a plant often leads to a reduction in leaf expansion. This observation has led to the suggestion that 'root factors' such as CKs are involved in the regulation of leaf growth. Indeed, the specificity of CKs in the promotion of cell expansion in radish cotyledons has formed the basis of a bioassay for this PGR. CK-stimulated growth of radish and cucumber cotyledons is associated with an increase in the plastic extensibility of the cell wall and the production of osmotically active sugars. The mechanism by which CK stimulates wall extensibility is unresolved; however, it has been demonstrated that the PGR enhances the K^+ uptake by leaf cells and that this event is crucial for accelerated elongation to be maintained.

The application of ABA to intact bean leaves inhibits light-stimulated cell enlargement and wall extensibility. ABA does not appear to prevent the light-stimulated acidification of the leaf surface, but reduces the capacity of the walls to undergo acid-induced wall loosening. Thus, *in vivo*, a stimulation of ABA biosynthesis as a result of water stress could ultimately lead to a reduction in leaf growth and a diminished area for transpiration. If ABA does play a role in leaf growth *in vivo*, the significance of this should be apparent from studies on ABA-deficient mutants. Paradoxically, ABA-deficient tomato mutants are characterized by reduced cotyledonary expansion, and the size of the cotyledon is correlated with the severity of deficiency of this PGR (Figure 8.1). Restricted cotyledonary development occurs under 100% RH, and is therefore not the result of low leaf turgor. It is therefore possible that whilst elevated ABA levels can inhibit cell extensibility, there is a requirement for a threshold level of ABA to maintain normal leaf expansion. Alternatively, ABA deficiency may have an indirect impact on the regulation of leaf cell growth.

A role for PGRs in fruit growth was first inferred from the classic studies carried out by Nitsch in the early 1950s on strawberries. His work elegantly demonstrated that achene removal inhibited strawberry receptacle enlargement, and that replacement of achenes with lanolin containing auxin or GA maintained fruit growth. Since achenes are a

Figure 8.1 ABA-deficient mutant tomato plants after 4 weeks of growth. The three plants on the right of the picture are double mutants. (Photograph courtesy of Dr Ian Taylor.)

rich source of these compounds, it was suggested that strawberry fruit growth is regulated by achene-derived PGRs. In a more recent investigation, Archbold and Dennis (1985) found that GA treatment could maintain receptacle growth of emasculated flowers for only 6 days, whereas growth of fruit treated with the synthetic auxin naphthalene acetic acid (NAA) continued for up to 30 days, albeit at a slower rate than that of pollinated flowers. In addition, achene removal up to 16 days after pollination stopped receptacle growth (Figure 8.2). The results of the study also demonstrated that removal of the achenes dramatically reduced the IAA content of the receptacle and the level of this PGR was elevated in parthenocarpic fruit by NAA treatment. These data support the hypothesis that auxins play a significant role in strawberry fruit growth; however, they do not preclude other PGRs from making a contribution. If GAs are involved, the receptacle may be only responsive to them early in development.

Evidence for a role of PGRs in the growth of other fruit is not as convincing as that in strawberry since the systems are not as amenable to experimentation. An alternative approach to surgical manipulation, is to induce parthenocarpic fruit and examine their endogenous PGR levels. GA_{4+7} is particularly effective at inducing parthenocarpy in species such as grape, apple, pear and tomato. GA-induced parthenocarpy in pears is associated with elevated levels of IAA and GAs in the receptacle, although these PGRs do not reach comparable levels to those found in pollinated plants. In tomato, GA-induced parthenocarpic fruit are not as large as seeded fruit, although they approach normal size when GA_3 is supplemented by auxin.

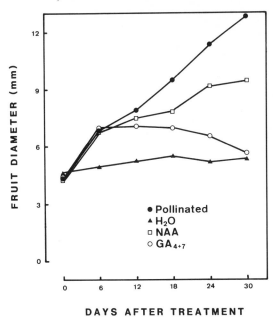

Figure 8.2 Growth of strawberry fruit after auxin (NAA) or GA$_{4+7}$ treatment. A single application of PGR was applied to emasculated flowers at the time of anthesis. (Redrawn from Archbold and Dennis, 1985.)

Although there is a tantalizing amount of information supporting a role for PGRs in leaf and fruit growth, convincing evidence is lacking. This is primarily because growth is such a broad developmental process that apart from events such as H^+ efflux or Ca^{2+} influx, other potential symptomatic changes have not been characterized. If some of the early mRNA changes which accompany auxin-stimulated hypocotyl growth are a prerequisite for shoot cell elongation *in vivo*, these changes may also be correlated with growth in tissues such as leaves and fruit. The influence of PGRs on these developmental changes could then be probed. In the absence of the identification of a key signal for leaf or fruit growth, a significant role for PGRs in these processes remains unproved.

8.2 Regulation of stomatal aperture

The primary function of stomata is to facilitate CO_2 and O_2 exchange between photosynthetic organs such as leaves and the environment. The nature of this role means that stomata function as pores through which water loss can occur, and therefore they act as valves which regulate the rate of transpiration. The relative importance of these two roles is dependent on environmental conditions, therefore a rapid and reversible mechanism must exist for the regulation of stomatal aperture.

It is now firmly established that the movement of K^+ is of crucial importance in the opening and closing of stomata. A range of stimuli can influence this process, including water stress, light and CO_2 levels. As a result of water stress, ABA levels in leaf tissues increase and stomata close. If ABA is applied through the leaf petiole, or to epidermal strips, rapid stomatal closure is elicited. For these reasons it is thought that the PGR plays an important role in the regulation of stomatal aperture, and this view has been strengthened by the discovery of putative ABA receptor proteins in guard cell protoplasts (see Chapter 9). In addition to water stress, ABA levels may be enhanced by waterlogging, salt stress, extremes of temperature, and by pathogenic attack, and all of these conditions can lead to stomatal closure. The intracellular distribution of ABA is probably determined by pH gradients, and the immediate response to water deficit may be a lowering of stromal pH in the mesophyll cells, resulting in an efflux of the PGR from the chloroplast. This phenomenon would occur prior to a stimulation in ABA biosynthesis, and could account for the common observation that stomatal closure precedes a detectable rise in ABA levels. The primary effect of ABA on K^+ movement is to stimulate efflux of this ion from the guard cells (see Figure 10.2), and this in turn raises their osmotic potential and the resultant loss of water precipitates stomatal closure. However, other ions may also be involved. For instance, ABA-induced stomatal closure is inhibited by the Ca^{2+} chelator EGTA or Ca^{2+} channel blockers (DeSilva *et al.*, 1985). Futhermore, the stomatal response to low concentrations of ABA is enhanced by the ion. These data imply that Ca^{2+} influx may be closely linked to the action of ABA in the regulation of stomatal aperture, and it has been proposed that the ion may act as a second messenger in the response (see Chapter 10).

ABA is not the only PGR which can affect stomatal aperture. In some species CKs can promote stomatal opening and antagonize the effects of ABA. The ability of CKs to maintain stomatal opening could contribute to the effects of water stress on stomata. If water deficit reduced the transport of CKs from the root to the shoot then stomatal closure would ensue. In support of this hypothesis, Davies *et al.* (1987) have reported that root drying reduces the CK level in maize leaves. IAA can also stimulate stomatal opening, however the efficacy of this PGR is dependent upon CO_2 levels, and in the absence of the gas the auxin has little effect.

It is evident that stomatal behaviour can be influenced by the application of PGRs. Since ABA-deficient mutants are characteristically 'wilty', and putative ABA receptors have been localized in guard cells, a role for this PGR in the regulation of stomatal aperture *in vivo* looks convincing. Both CKs and IAA may also be involved, but the evidence remains circumstantial. Interestingly, stomata from adaxial and abaxial leaf surfaces exhibit differential sensitivities to ABA, and this phenomenon is maintained in peeled epidermal layers (Figure 8.3). Using photoaffinity labelling techniques, it may be possible to quantify the number of putative ABA receptors in guard cells from adaxial and abaxial leaf surfaces. The results of such a study could help to pinpoint the cellular basis for modulated PGR sensitivity in plant tissues.

8.3 Epinasty and hyponasty

The orientation of leaves within a canopy is important in order to maximize interception of solar radiation, and minimize shading of

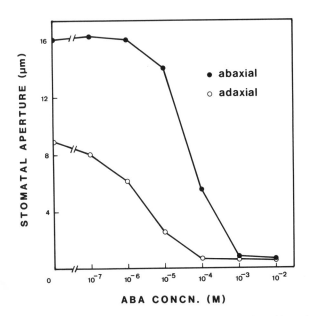

Figure 8.3 Differential sensitivity of stomata from adaxial or abaxial epidermal strips from leaves of *Commelina communis* to ABA. (Redrawn from De Silva *et al.*, 1986.)

photosynthetically active tissues. From time to time, other constraints may also be imposed on a canopy, and in particular it has been proposed that heat load on the leaf surface may be reduced under conditions of stress by reorientation of the lamina. One mechanism which may achieve this is epinastic bending. Nastic curvature can be induced by a range of non-directional stimuli, and the angle of bending is determined by the morphology of the tissue. The epinastic bending exhibited by leaf petioles from dicotyledonous plants is one of the most common examples of such a response. Bending is achieved by the adaxial surface of the petiole elongating more rapidly than the abaxial, and the leaf flexes downwards.

Leaf epinasty can be induced by such stress conditions as root waterlogging or anoxia, pathogenic attack, and physical restriction. In addition, these conditions can lead to accelerated leaf senescence, and premature leaf and flower abscission. Since these phenomena are symptomatic of ethylene treatment, and the gas is known to induce leaf epinasty, it has been suggested that the PGR plays an important role in this developmental event *in vivo*. Further evidence in favour of this hypothesis is the demonstration that leaf epinasty in tomato plants can be prevented by AVG pretreatment prior to exposing the roots to anaerobic conditions.

Some progress has been made in establishing the chain of events leading to leaf epinasty in waterlogged tomato plants. It has been shown that the low O_2 content of waterlogged soil stimulates the production of a root factor which moves into the shoot system via the xylem where it promotes epinasty. This factor has been identified as ACC; this compound is known to accumulate under conditions of anaerobiosis and be converted to ethylene when O_2 becomes available (Bradford and Yang, 1980). The simplest mechanism to account for ethylene-induced epinasty is that the adaxial and abaxial surfaces of the petiole are differentially sensitive to the gas. In *Helianthus annuus*, ethylene transiently promotes growth of the upper surface and inhibits it on the lower side. Other PGRs such as IAA can also induce epinastic curvature; however, this can be prevented by AVG or Ag^+, which implies that auxin-stimulated ethylene is responsible. Environmental stimuli such as gravity can also induce leaf epinasty, but the mechanism by which this is mediated is unknown.

In young leaves, the epinastic curvature is reversible. The regulation of this hyponastic straightening has not been resolved, although it has been proposed that it is due to the release of growth potential on the lower side after the elevated ethylene level has subsided. Alternatively, the exhaustion of a growth factor on the upper side of a petiole may occur more rapidly than on the lower. GAs have been reported to induce hyponasty in plagiotropic organs, but the mode of operation of this PGR has not been investigated in detail.

In contrast to tropic bending, epinastic and hyponastic curvatures are apparently associated with a uniform PGR distribution. This implies that a differential sensitivity to a PGR mediates the response. In Chapter 4, possible scenarios underlying sensitivity differences were described. Petiole epinasty would seem to be a further system in which to probe the mechanism of these differences.

8.4 Ripening

After fertilization, the role of the fruit is to protect and nurture the developing seeds. Once maturation has been reached, the structure of the fruit alters significantly in order to expedite seed dispersal. The changes that take place in a fleshy fruit render it conspicuous and palatable to a potential consumer, and together constitute the syndrome of ripening. A glance at the diversity of fruit types reveals that there are a variety of ways by which this can be accomplished. The biochemical pathways which are implemented to give a fruit its characteristic colour, texture, flavour, and aroma, must vary from species to species; however, it is evident that they represent a highly co-ordinated sequence of events. Ripening, therefore, is not a degenerative process arising out of cell autolysis, but an example of a precisely regulated developmental programme (Brady, 1987).

The adage that an apple will ripen an unripe banana was the starting point in the elucidation of the role of PGRs in fruit ripening. This observation was recorded by Cousins in 1910, and 24 years later Gane provided the first chemical evidence that fruit produced ethylene. Fruits can be classified as climacteric or non-climacteric on the basis of whether or not they exhibit a respiratory climacteric during ripening (Table 8.1). Climateric fruit can be induced to ripen by exposure to ethylene, while non-climacterics exhibit little response to the gas unless the treatment is prolonged extensively. A further feature which distinguishes climacterics from non-climacterics is that ripening in the former group is accompanied by a rise in ethylene production. In tomato, cantaloupe, banana and avocado fruit, this stimulation in ethylene biosynthesis is a result of an increase in the activity of both ACC synthase and the

Table 8.1 Classification of fruits on the basis of their respiratory behaviour during ripening.

Climacteric fruits

Apple (*Malus sylvestris*)
Avocado (*Persea americana*)
Banana (*Musa* sp.)
Chinese gooseberry (*Actinidia chinensis*)
Mango (*Mangifera indica*)
Cantaloupe melon (*Cucumis melo*)
Papaya (*Carica papaya*)
Peach (*Prunus persica*)
Pear (*Pyrus communis*)
Plum (*Prunus* sp.)
Tomato (*Lycopersicon esculentum*)

Non-climacteric fruits

Sweet cherry (*Prunus avium*)
Sour cherry (*Prunus cerasus*)
Grape (*Vitis vinifera*)
Grapefruit (*Citrus paradisi*)
Lemon (*Citrus limonia*)
Orange (*Citrus sinensis*)
Pineapple (*Ananas comosus*)
Strawberry (*Fragaria ananassa*)

ethylene-forming enzyme (EFE) (Figure 8.4). In climacterics, the rise in ethylene is one of the first signs of ripening, and is thought to be sustained by autocatalytic stimulation. No autocatalytic stimulation of ethylene can be induced in non-climacterics, although application of the gas does promote the respiration of the fruit as long as the PGR is present. Thus non-climacteric fruits have the ability to perceive ethylene although in general the gas does not stimulate their ripening (Roberts *et al.*, 1987).

The sensitivity of many climacteric fruits to ethylene is variable both between and within species. For example, morphologically similar mature green tomato fruits may require a different number of days' exposure to ethylene before they will show signs of ripening. The duration of the lag phase is perhaps dependent upon their 'physiological' age, and may reflect a decline in the level of a ripening inhibitor to below a critical threshold. Further evidence in favour of the existence of an inhibitor is the demonstration that detachment of many climacteric fruit hastens their ripening. Indeed, fruits such as avocado will *only* ripen

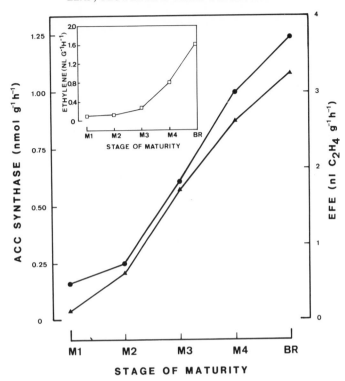

Figure 8.4 Activity of ACC synthase and EFE (ethylene-forming enzyme) in locule tissue from tomato fruit at different stages of development. M_1, immature-green; M_2, partially mature-green; M_3, mature-green; M_4, advanced mature-green; BR, breaker (fruit showing the first signs of ripening). Inset: ethylene production by fruit of equivalent developmental stages. (Modified from Brecht, 1987.)

after detachment from the intact plant. As yet, no specific ripening inhibitor has been isolated, although the process can be retarded by the application of either IAA or GAs. The effects of applied auxin are complicated by the capacity of this PGR to promote ethylene biosynthesis. Ripening is delayed if fruits are vacuum-infiltrated with IAA so that the auxin becomes distributed evenly throughout the fruit; otherwise there is a tendency for the enhanced ethylene production induced by the auxin treatment to stimulate the ripening proces. Recently it has been demonstrated that NAA and 2,4-D will inhibit the ripening of de-achened strawberry fruit. This observation suggests that auxins could play a strategic role in regulating the ripening of non-climacteric fruit.

Application of GAs to orange or tomato fruits retards their chlorophyll loss, and specifically inhibits softening in the latter tissue. Benzyladenine has also been reported to inhibit chlorophyll loss in some fruit. The effects of this CK on degreening may be a reflection of its ability to inhibit senescence, rather than ripening events *per se*.

The biochemical changes that precipitate ripening in climacteric fruit include increases in the activity of enzymes responsible for ethylene biosynthesis, cell wall degradation, and pigment accumulation. Studies on the molecular biology of ripening in avocado and tomato indicate that these changes are the result of selective gene expression. In both these systems, the evidence is overwhelming that ethylene regulates the appearance of a group of ripening-related mRNAs. In avocado, one of the mRNAs has been identified to code for the cell wall degrading enzyme β-1,4-glucanase (Tucker *et al.*, 1985), while in tomato, an mRNA for the softening enzyme polygalacturonase (PG) has been characterized. The gene for PG in tomato fruit has been sequenced (Grierson *et al.*, 1986), and the cDNA coding for it used to quantify PG mRNA levels. The results demonstrate that ethylene can rapidly stimulate the appearance of PG mRNA in tomato. Both ripening and the increase in PG mRNA can be retarded by Ag^+. Furthermore, the ripening mutant *rin* shows no accumulation of PG mRNA and does not soften (see Figure 10.4). Thus ethylene specifically stimulates the appearance of mRNA coding for an enzyme which plays a crucial role in tomato fruit ripening. A cDNA library constructed using mRNA from ripe tomato fruit has revealed that over 20 mRNAs appear during ripening, and there is evidence that a number of mRNAs disappear or decline significantly. One of the mRNAs which increases in concert with ripening is also elevated by wounding, and may code for a protein associated with ethylene biosynthesis or metabolism. The identity of the proteins which are encoded by the other mRNAs remains to be elucidated. Although it is tempting to speculate that the appearance of the ripening-related mRNAs is under transcriptional control, this awaits verification.

The recent progress that has been made to pinpoint the molecular events which accompany fruit ripening, has enabled the role of PGRs in the process to be examined more critically. The results from studies on avocado and tomato demonstrate convincingly that ethylene is an integral component in the regulation of ripening of climacteric fruit (see Figure 8.5). So far, the systems have not been utilized to examine whether the inhibitory properties of IAA, GA or CK have any significance in the control of ripening *in vivo*. Neither has the phenomenon of sensitivity been probed with the aid of molecular tools. With regard

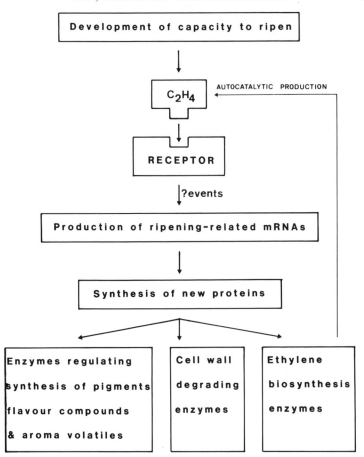

Figure 8.5 Scheme of climacteric fruit ripening. (Modified from Grierson, 1986.)

to non-climacterics, the identity of the ripening regulator *in vivo* remains a mystery. Whether the current band of PGRs plays a role in ripening of such fruit awaits the study of the molecular biology of ripening of a non-climacteric fruit.

8.5 Senescence

Senescence is the sequence of biochemical and physiological events which culminate in cell death. During the life cycle of a plant, cell senescence may be restricted to individuals such as xylem or root cap cells, occur synchronously in an organ such as a leaf, flower, or fruit, or proceed throughout an entire organism as in the senescence of monocarpic plants. Whilst the process ultimately leads to the degeneration of the internal structure of a cell, studies of senescence indicate that disassembly does not take place randomly, but, like ripening, proceeds in a highly coordinated fashion. The fact that the metabolic changes which accompany senescence are regulated implies that the process is mediated by specific chemical signals.

The simplest system on which to study senescence is that of isolated leaves. Excision of this tissue accelerates senescence and removes potential interactions from the rest of the plant. In addition, excised leaves can be exposed conveniently to a range of different chemicals, and maintained under a variety of environmental stimuli. Whether the developmental programme induced under these conditions represents an accurate reflection of that which occurs during 'natural' senescence is a matter for debate.

Senescence of excised leaves can be arrested if roots are allowed to develop at the cut surface of the petioles. This demonstration was reported by Chibnall in 1939, and remains the basis for the hypothesis that roots contain a senescence-retarding compound. Attempts to identify potential candidates for this role have relied on the application of compounds such as PGRs to excised leaf tissues. The results obtained vary considerably between species. In *Xanthium* and many other species, added CKs retard leaf senescence, and localized application of kinetin to tobacco leaves has been shown to maintain the level of chlorophyll in the area of treatment while the rest of the leaf undergoes chlorosis. Senescence of *Taraxacum officinale* leaves cannot be delayed by CKs, but the process is inhibited by GAs. In *Rumex* species and nasturtium, GAs and CKs are both effective in retarding senescence. Some reports have indicated that auxins may also have the capacity to delay senescence, but in general this is only evident at high concentrations. Senescence promoters can also be assayed using the excised leaf system, and ABA, ethylene, and the compound jasmonic acid have been found to be effective in hastening the process. Attempts have been made to correlate the effects of PGRs on leaf senescence with their abilities to regulate stomatal aperture. This has led to the proposal that CKs delay senescence by virtue of their capacity to maintain stomata in an open position (see section 8.2), thus ensuring that movement of solutes to the leaf tissue is sustained, and that potential senescence-promoting volatiles such as ethylene or methyl jasmonate do not accumulate. This proposal envisages that ABA promotes leaf senescence through its ability to close stomata. Such an hypothesis may prove to be a over-simplification of the situation, particularly in view of Thimann's recent demonstration (1985) that the senescence of excised nasturtium leaves maintained in the dark is not directly correlated with stomatal aperture.

Although there are reports of a correlation between leaf senescence and levels of auxins, GAs, CKs, ABA, or ethylene, in general no consistent pattern has been established. Moreover, the use of correlative data has severe limitations. One approach which has been used to probe the role of ethylene in leaf senescence more critically is the application of AVG or Ag^+ to plant tissues. The results of such studies demonstrate that blocking of either ethylene biosynthesis, or action, will retard the rate of chlorophyll loss of both tobacco leaf discs and excised oat leaves, and imply that endogenous ethylene may be involved in the senescence programme in some way (Roberts et al., 1985). It has been suggested that the ability of polyamines to delay the senescence of excised leaf tissue may be due to their capacity to inhibit the conversion of ACC to ethylene. Alternatively, the effects of polyamines could be attributable to their tendency to scavenge free radicals and hence alleviate the effects of the peroxidative reactions which are an inherent feature of senescence.

Leaf senescence is characterized by chlorophyll loss and extensive degradation of nucleic acids and protein. However, the phenomenon is also associated with the synthesis of protein and RNA. Since cyclohexi-mide inhibits leaf senescence, the synthesis of new proteins must be crucial for the process to proceed. In addition, quantitative and qualitative changes in mRNA populations have been demonstrated to occur during the senescence of excised wheat leaves and soybean cotyledons. In the light of these observations, it would seem that leaf senescence must be accompanied by changes in gene expression, it would therefore seem to be an ideal system to probe using molecular techniques. For instance, cDNA libraries generated from mRNA extracted from green and senescing leaves could be employed to highlight senescence-related messages, and once these have been characterized, the regulation of their expression by specific PGRs could be assessed.

It is well established that ethylene is involved in the senescence of flowers. In carnations, petal wilting, which is the first visible sign of senescence, is preceded by a rise in ethylene and respiration. If this rise in ethylene is prevented by AVG treatment, or the action of ethylene is inhibited by Ag^+, flower senescence is delayed. This latter treatment

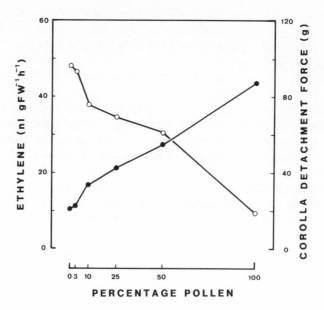

Figure 8.6 Response of *Digitalis purpurea* flowers 18h after pollination with mixtures of pollen and powdered glass (w/w). Both the rate of ethylene production ● and the force required to detach the corolla ○ are proportional to the amount of pollen applied. (Redrawn from Stead, 1985.)

forms the basis of a commercial treatment to prolong the vase life of carnation blooms in the horticultural industry (see Chapter 11). The ethylene climacteric produced by the flower is initiated *in vivo* by pollination, and in carnation is precipitated by an increase in the activity of ACC synthase and EFE activity. The nature of the elicitor that stimulates ethylene biosynthesis is unknown; however, the mechanism of its generation must be very sensitive, since in *Digitalis purpurea* it responds quantitatively to the amount of pollen applied (Figure 8.6). It has been suggested that ethylene plays a less important role in the senescence of ephemeral flowers such as *Ipomea* and *Hibiscus* species since an increase in production of the gas can only be detected after the onset of flower fading. However, since Ag^+ retards senescence of *Hibiscus* petals if applied prior to flower opening, it is possible that a change in sensitivity to ethylene results in the petals responding to low preclimacteric levels of the gas. The sensitivity of this tissue to ethylene has been shown to increase dramatically on the day of flower opening.

Senescence of monocarpic plants such as cereals and soybean is dictated by the time of fruit maturation. If the developing fruits are removed the life of the plant is prolonged. Excision of the apex of soybean plants generates Y-shaped individuals, and these have been employed to investigate the existence of a senescence promoter in monocarpic plants. If fruits are removed from one branch only, leaves on the depodded branch remain green while those on the other branch senesce. Although this may indicate that developing seeds are the source of a 'senescence factor', conclusions from this work should be made cautiously (for a detailed discussion, see Sexton and Woolhouse, 1984). Certainly, progress on the isolation of a putative senescence factor has been slow. Although ABA levels increase in maturing soybean fruits, and this PGR will accelerate senescence of intact plants, ^{14}C-ABA moves preferentially from the leaves to the fruit, which indicates that it is unlikely to be the senescence factor. The timing of senescence in monocarpic plants may also be dependent on the levels of senescence-retarding compounds, and a synergistic effect between GAs and CKs in the delay of senescence has been discovered in soybean explants.

In conclusion, although PGRs have been implicated in the senescence of leaves, flowers and whole plants, in the absence of a thorough knowledge of the biochemistry of these processes, it is difficult to assess the significance of their role. The evidence in favour of a role for ethylene in flower senescence is more convincing; however, it remains to be determined whether the PGR acts as a critical regulator of the senescence of other cells and tissues.

8.6 Abscission

Abscission is the process by which plant parts are shed. It occurs at discrete sites in a plant and is mediated by the dissolution of the wall between adjacent cells within the separation zone. Abscission is not restricted to the shedding of leaves, flowers or fruit but is also responsible for the loss of many other aerial parts of a plant. The site at which abscission takes place is commonly distinguishable morphologically prior to cell separation. It is composed of a row (or rows) of cells which are more compressed than their neighbours. During separation the cytoplasm of these cells shows signs of secretory activity with a proliferation of endoplasmic reticulum and Golgi structures. The cell wall then dilates and finally the middle lamella is degraded (Figure 8.7).

The initial demonstration that ethylene could promote abscission can

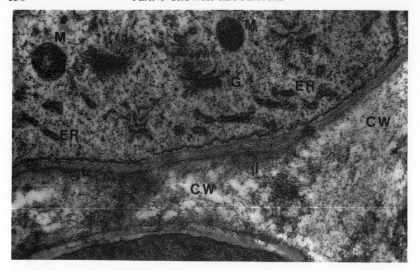

Figure 8.7 Electron micrograph of abscission zone cells from rape flowers separating in response to ethylene. *CW*, cell wall (exhibiting extensive degradation); *ER*, endoplasmic reticulum; *G*, Golgi structure; *M*, mitochondrion. (Photograph courtesy of Dr Paul Meakin.)

be traced back to Girardin's observation (1864) that leaking 'illuminating gas' caused the defoliation of the surrounding vegetation. Since then, numerous reports that ethylene can stimulate organ shedding have been recorded, and it has become universally accepted that the gas plays a prominent role in the natural regulation of abscission. *In situ*, the timing of shedding is regulated by the ethylene which emanates from the tissues distal to the site of separation, and production of the gas increases during the progress of senescence and ripening. Not all plant organs are equally sensitive to applied ethylene. For instance, tomato flowers will abscind within 6 h of exposure to the PGR, while leaves or fruit from the same species may take over 48 h before they exhibit signs of separation, depending on their physiological age and condition. Length of the lag period prior to separation can be curtailed by excision of the tissue distal to the abscission zone, which suggests that a basipetally transported abscission inhibitor may also be involved in the process. Progress on the identification of an abscission-retarding compound was first made by Laibach in 1933. He applied detached orchid pollinia rich in auxin to the stump of a debladed petiole and found that abscission was delayed. Further work has conclusively shown that auxin is a potent abscission retarder, and that if applied to tissues early enough it will counteract the abscission-promoting properties of ethylene. Studies on leaf abscission have revealed that both the level of IAA in the lamina and the efficacy of the auxin transport system decline prior to 'natural' abscission (see Chapter 4). Observations such as these have led to the formulation of the hypothesis that the IAA: ethylene status of the tissue is the crucial determinant in regulating the onset of abscission (Sexton and Roberts, 1982).

The name 'abscisic acid' was derived from the early demonstration that the PGR could accelerate the abscission of young cotton fruit. Paradoxically, ABA is not particularly effective at stimulating organ shedding when applied to intact plants, although it has a potent effect on excised abscission zones or 'explants'. Such isolated zones are routinely employed in studies on abscission, since their generation synchronizes the onset of the process which gives rise to a reproducible system on which to work. An explanation for the ability of ABA to promote explant abscission may lie in the capacity of this PGR to accelerate senescence of excised tissues. When applied distally to *Phaseolus vulgaris* explants, ABA brings forward the ethylene-climacteric associated with the senescence of the pulvinar tissue. Suppression of the ethylene climacteric by AVG, minimizes the abscission-promoting effects of ABA. If ABA hastens abscission solely by stimulating ethylene production, then it should be ineffective in the presence of saturating levels of the gas, or under hypobaric pressures where the gas is rapidly removed from the tissues. Since there are reports of ABA retaining its abscission-accelerating properties under these conditions, it remains a possibility that this PGR could have a direct effect on cell separation. The conclusion from this discussion is that the timing of abscission could be attributable to a change in level of at least three PGRs, and furthermore, that these PGRs have the capacity to interact with one another. In an attempt to rationalize this, Sexton *et al.* (1985) have developed a scheme that places the regulation of the timing of abscission under multifactorial control. This useful model may provide an insight into the complexities of developmental regulation.

Associated with the onset of cell separation is a rise in the activity of a number of hydrolytic enzymes. These include, β-1,4-glucanase (cellulase) and polygalacturonase (PG). The increase in cellulase has been attributed to the *de novo* synthesis of a particular isoenzyme with an isoelectric point of 9.2. The activity of this enzyme is promoted by ethylene and inhibited by IAA. The pI 9.2 enzyme has been purified from

Phaseolus vulgaris abscission zones, and antibodies raised against it. Peroxidase-linked antibodies have been shown to bind specifically to ethylene-treated bean abscission zones. Furthermore, if the antibody is injected into the abscission zone of bean, cell separation is prevented. Although this work provides convincing evidence that cellulase has a key role in abscission, it does not shed light on why the pectin-rich middle lamella is the first area of the cell wall to exhibit signs of breakdown prior to separation. Recently, a number of reports have appeared correlating PG activity with abscission. Application of a purified preparation of this enzyme has been shown to degrade the middle lamella of tomato fruit. It is likely, therefore, that both PG and cellulase make a significant contribution to cell separation.

The abscission of explants is prevented by cycloheximide. Actinomycin D will also inhibit cell separation, although it is ineffective in such systems as flower abscission where the lag period in response to ethylene is less than 6 h. It has recently been shown in *Phaseolus vulgaris* that ethylene enhances the levels of several translatable mRNA species during leaf abscission (Kelly *et al.*, 1987). The identity of the proteins which these messages encode is unknown. This is as far as the molecular biology of abscission has progressed. In many respects this is surprising, in view of the marked similarities that this phenomenon has with ripening and the advances that have been made in our understanding of this latter process. One explanation for the slow progress on abscission is that the separation zone represents only a few rows of cells. Pursuing the traditional molecular path of enzyme and mRNA isolation is therefore difficult and time-consuming. Since some of the events which occur during ripening show marked similarities to those that take place during abscission, one way to 'short-circuit' the procedure might be to utilize the molecular probes developed in the ripening systems to study abscission. This approach could rapidly give us information on whether abscission is truly under multifactorial control, or whether ethylene and IAA are the principal regulatory PGRs.

8.7 Conclusions

With the aid of techniques of molecular biology it is now possible to identify some of the earliest changes which are associated with developmental events such as growth, ripening, senescence and abscission, and therefore critically probe the effect of PGRs on these changes. By this approach we are slowly reaching a position where we may be able to assemble enough pieces of the jigsaw to arrive at a firm conclusion as to whether a PGR regulates a particular developmental process. We are well on the way to achieving this in the ripening system, and it is only a matter of time before we make similar progress in senescence and abscission. In developmental processes such as apical dominance and bud dormancy, although there are signs that these events may be amenable to the molecular approach, this has yet to be proven. The tropic and nastic responses described are unlikely to be mediated by differential gene expression, and therefore the novel techniques of molecular biology will not directly come to our aid. In these systems further progress may come from the study of developmental mutants. Many mutants have now been characterized in *Arabidopsis*, tomato, pea and maize, and these include a number of mutants with aberrant tropic phenotypes. By comparing and contrasting mutants with normal plants it should be possible to focus on the strategic events responsible for tropic and nastic bending and to examine the effect of PGRs on these changes. The usefulness of mutants is not restricted to these phenomena but is also yielding valuable information about the regulation of growth, ripening, senescence, abscission and many other biochemical and physiological events (Thomas and Grierson, 1987).

A further approach that may prove enlightening is to study the impact of specific DNA constructs on the phenotype of transgenic plants. Some success has already been achieved in this area with the demonstration that petunia plants which have been transformed with two tDNA auxin biosynthetic gene products from *Agrobacterium tumefaciens* exhibit strong apical dominance (Klee *et al.*, 1987).

It is evident from the previous chapters that a change in the sensitivity of a tissue to a specific PGR is commonplace. Although there is increasing interest in the underlying basis of this phenomenon, as yet no detailed studies have been undertaken to investigate this. The stomatal system might be particularly amenable to study, since differential sensitivity to ABA of abaxial and adaxial surfaces is maintained in epidermal strips. Furthermore, the techniques are potentially available using photoaffinity labelling to quantify receptor number and affinity. The results of investigations such as this are awaited with keen interest.

CHAPTER NINE

RECEPTORS — SITES OF PERCEPTION OR DECEPTION?

Plant cells have the capacity to distinguish PGRs from the myriad of other molecules to which they are exposed. In addition, they can discriminate between naturally occurring PGRs and synthetic derivatives having subtly different chemical structures. These observations provide compelling evidence in favour of the existence of specific binding sites for PGRs. The stringent structure–activity relationships described demand that the molecular recognition of these sites must be precise, and this feature is most compatible with the receptor being a protein. Hormone receptor proteins have been isolated and characterized from animal cells, and their discovery has been frequently cited in support of the hypothesis that similar sites must exist in plants. The accuracy of this analogy remains to be seen.

9.1 Binding studies

9.1.1 Theory

Using our knowledge of the way in which plant tissues respond to a change in concentration of PGRs or their analogues, certain predictions can be made about the properties which a putative receptor would be expected to exhibit. The binding should be:

(i)   *Reversible* — since the rate of many developmental responses slows significantly when the PGR is removed
(ii) *Of high affinity* — since endogenous levels of PGRs are very low
(iii) *Saturable* — the range of PGR concentrations over which binding increases and finally saturates should be compatible with the concentration range over which the response to the PGR saturates. However, from our knowledge of animal hormone receptors, this may not always be true. Sometimes maximal response may be achieved at a receptor occupancy which is substantially less than 100%.
(iv) *Specific* — binding by different analogues of a PGR should correlate with their biological activities and should also be restricted to

a particular class of PGR. However, if the biological activity of an analogue is primarily regulated by its uptake or transport this may not be reflected in a binding assay.

(v) *Confined* to tissue having a 'response' to the PGR

(vi) *Linked* to a biological response.

On the basis of these criteria, binding experiments are performed to determine the presence or absence of a putative receptor in an extract from a tissue. Criteria (i)–(iv) are usually investigated. Criterion (v) is less often addressed, and (vi), although of greatest significance, is the most difficult to establish and therefore rarely studied. The data from binding assays can be analysed to yield two important binding constants. These are K_d, the dissociation constant, which is a measure of binding affinity; and n, the number of binding sites associated with the tissue.

If L represents a ligand (the PGR), R a receptor and $L-R$ the ligand–receptor complex, then, from the Law of Mass Action:

$$L + R = L-R$$

It can be calculated from this relationship that at equilibrium, the concentration of bound ligand B is given by

$$B = \frac{nK_aF}{1 + K_aF}$$

where F = concentration of free or unbound ligand, and
K_a = association constant,

or

$$B = \frac{nF}{K_d + F}$$

where n = number of binding sites and
K_d = dissociation constant ($1/K_a$).

To present the binding data graphically, the following linear transformations are used:

$$\frac{1}{B} = \frac{K_d}{nF} + \frac{1}{n}$$

(double reciprocal plot of $1/B$ v. $1/F$)

and

$$\frac{B}{F} = \frac{n}{K_d} - \frac{B}{K_d}$$

(Scatchard plot of B/F $v.$ B).

The binding constants K_d and n can be obtained from the slope or intercept of these plots, respectively (Figure 9.1). These methods of analysis can only cope with simple binding reactions. Other graphical methods can be utilized to accommodate multiple binding sites with different affinities and co-operative interactions.

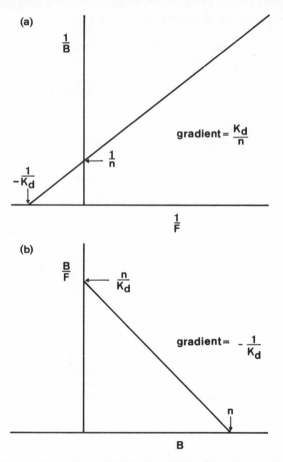

Figure 9.1 Linear transformations of binding data. (a) Double reciprocal plot; (b) Scatchard plot.

In order to estimate K_d and n, the values of B (the concentration of bound PGR) and F (the concentration of free PGR) must be measured. These values must be determined over a range of total PGR concentrations which fall below receptor saturation. This is most readily achieved by the use of radiolabelled PGR with high specific activity. The appropriate range of concentrations is obtained by the addition of increasing amounts of unlabelled PGR to a small quantity of radioactive PGR. The specificity of a receptor can be determined by measuring the ability of an analogue to displace labelled PGR from the binding site. Both soluble and membrane-bound proteins can be assayed for binding activity in this way, although the experimental protocols vary.

9.1.2 Practice

Binding of a PGR to a soluble or solubilized receptor can be assessed by a number of methods. The most reliable are:

(i) *Equilibrium dialysis.* The extract is dialysed to equilibrium against labelled PGR. Dialysis cells are available commercially for multiple assays. The value of F is given by the amount of PGR in the non-protein half of the cell, and B is equal to the difference in radioactivity between the two cell halves (Figure 9.2).

(ii) *Gel filtration.* Separation of bound and free PGR is achieved by gel filtration through a Sephadex column. Proteins elute at the void volume and any radioactive PGR that binds to the column can be quantified. Free PGR elutes later. For rapidly dissociating complexes, gel filtration can be carried out under equilibrium conditions by equilibrating and eluting the column with the same concentration of labelled PGR as in the sample.

(iii) *Ultracentrifugation.* Protein–PGR complexes can be separated from free PGR by ultracentrifugation through a density gradient. This technique can also be performed under equilibrium conditions.

A number of other methods have been employed, including selective adsorption of free PGRs by charcoal; immobilization of bound PGRs to nitrocellulose filters; and ultrafiltration of PGR plus receptor mixture. Assays relying on precipitation of the protein–ligand complex by ammonium sulphate have been criticized, since 'artificial' binding sites can be induced by this procedure. The relative merits of these various assay methods are discussed by Venis (1985).

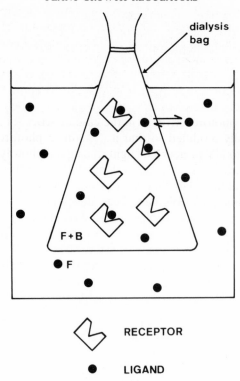

RECEPTOR

LIGAND

Figure 9.2 Affinity determination by equilibrium dialysis. The extract containing the putative receptor is dialysed against radiolabelled ligand (PGR). The concentration of free ligand (F) is equal to the amount of PGR outside the dialysis cell and the concentration of bound ligand (B) is equivalent to the difference in radioactivity between that in the cell and that in the surrounding solution.

Free and bound PGR concentrations are relatively easy to quantify when particulate, e.g. membrane-bound, proteins are assayed. The bound ligand can be readily separated from the free by centrifugation or filtration methods.

A major problem in the study of PGR binding systems is the reversible nature of the ligand–receptor interaction. Affinity labelling aims to overcome this problem by covalently linking a radioactive or fluorescent ligand to the receptor. This can be done *in vivo* with plant protoplasts, or *in vitro* with subcellular fractions or purified receptor preparations. The ability to label a receptor *in vivo* is extremely useful because it eliminates the risk of artefacts arising as a result of the extraction procedures. The covalently linked ligand–receptor complex can be subjected to rigorous purification or separation procedures. There are two types of affinity labelling, chemical or photoaffinity (Figure 9.3). In the former, the ligand is modified to contain a reactive group that attaches to the binding site of the receptor. The drawback of this technique is that attachment commonly occurs outside the receptor binding site or nonspecifically to other molecules. In photoaffinity labelling, the ligand may already be photoreactive, as in the case of ABA for example, or it can be chemically modified so that it contains a photoreactive group. When exposed to UV or near-UV light, the ligand photolyses to a highly reactive intermediate capable of covalently linking to the receptor binding site. Photoaffinity labelling is also not without nonspecific binding problems, but protocols have been developed to minimize these. Affinity labelling not only enables potential receptors to be identified in a mixed population of proteins, but it can also provide valuable information about the molecular structure of a binding site from the sequence of labelled receptor peptide fragments. Such techniques have been applied to study animal receptor systems with considerable success, and the reader is referred to Ruoho *et al.* (1984) and Gronemeyer (1985) for further details.

9.2 Binding sites for PGRs

9.2.1 Auxins

Auxin-binding proteins are the best-known receptor system in plants (see Libbenga *et al.*, 1985). Several classes of auxin-binding proteins have been identified and partially characterized. These comprise binding sites for auxin transport inhibitors, membrane-bound auxin-binding sites, and soluble/nuclear bound auxin-binding sites. Each of these classes will be considered in turn.

The cellular location of PGRs can have a significant impact on their role during differentiation and development. Auxins are transported by specific carriers (see Chapter 4), one of which has been characterized by its ability to bind the synthetic auxin transport inhibitor N-1-naphthylphthalamic acid (NPA). The NPA binding protein was first detected in maize membrane preparations and has subsequently been found in a variety of plant species. Binding is reversible and of high affinity, with values of K_d varying between different plant species. In membranes from tobacco cell suspensions, a value of K_d as high as 3×10^{-9}M has

CHEMICAL–AFFINITY LABELLING

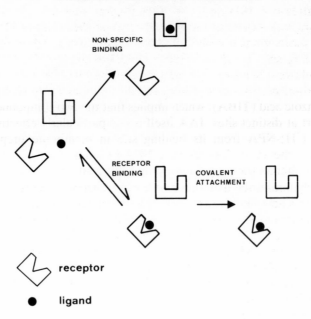

NON-SPECIFIC BINDING

RECEPTOR BINDING

COVALENT ATTACHMENT

receptor

ligand

PHOTO–AFFINITY LABELLING

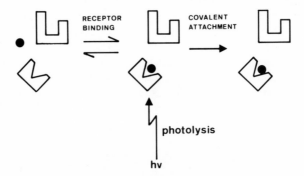

RECEPTOR BINDING

COVALENT ATTACHMENT

photolysis

hv

Figure 9.3 Chemical affinity and photoaffinity labelling. In the former (*upper*), the ligand (PGR) is modified so that it contains a reactive group that can attach to a functional group at the binding site of the receptor. Under unfavourable conditions, attachment can occur outside the receptor binding site or nonspecifically to other molecules. In photoaffinity labelling (*lower*), the ligand may already be photoreactive (as in the case of ABA), or it can be modified chemically to generate a photoreactive group. When exposed to UV light, the ligand photolyses yielding a highly reactive intermediate which has the capacity to form a covalent linkage with the receptor site.

been reported. Values of n vary too, although on average there are approximately 6×10^3 to 6×10^4 NPA binding sites per cell. There is a good correlation between binding affinity and the effect of NPA on auxin transport. Furthermore, NPA derivatives displace (^{3}H)-NPA from the binding site in a manner compatible with their activities as transport inhibitors. However, the binding is not displaced by physiologically active concentrations of another inhibitor of polar auxin transport, triiodobenzoic acid (TIBA), which implies that the two compounds inhibit transport at distinct sites. IAA itself is not particularly effective at displacing (^{3}H)-NPA from its binding site in membrane preparations. However, the auxin does displace NPA in binding protein solubilized from the membrane preparation with detergent.

The chemiosmotic hypothesis of auxin transport proposes that polar transport is achieved by carrier-mediated auxin anion efflux through the plasmalemma at the basal ends of transporting cells. Auxin transport inhibitors are thought to block this efflux carrier. Using monoclonal antibodies to the NPA binding protein, the receptor has been visualized by immuno-light microscopy (see Figure 4.5) and found to be associated with the plasma membrane at the basal region of cells adjacent to the vascular bundles (Jacobs and Short, 1986). This is the first localization of a receptor/carrier molecule in plant cells.

Maize coleoptile membranes contain other binding proteins with properties more consistent with those expected of auxin receptors (Table 9.1). Similar proteins appear to exist in other monocotyledons, but have not yet been demonstrated in dicots. Three auxin binding sites, design-

Table 9.1 Characteristics of membrane-bound auxin binding proteins in maize

Nomenclature	K_d for NAA ($\times 10^{-6}$M)	pmol g^{-1}	Cellular location	Reference
Site 1	0.15	25–30	ER/Golgi	Batt et al. (1976)
Site 2	1.60	100–120	Plasmamembrane	
Site I	0.4	40	ER	Dohrmann et al. (1978)
Site II	1.3	20	Tonoplast	
Site III	5.0	40	Plasmamembrane	
—	0.5–0.7	30–50	ER/Golgi Plasmamembrane	Ray et al. (1977)

For further details and references see Venis (1985).

ated I, II and III, have been described. There is, however, some dispute about the distinction between I and II.

Site III binding has been characterized in maize and zucchini membrane preparations. It has been suggested that site III is a plasma membrane-bound auxin transport protein, and that 'binding' to this site represents the net accumulation of radiolabelled auxin into right-side-out plasmalemma vesicles. These vesicles are fragile and sensitive to detergent, and 'binding' can be disrupted by freeze-thawing and sonication. In contrast, radiolabelled IAA bound to sites I and II of maize membrane preparations is not released by detergent treatment and appears to represent genuine binding of the PGR to a membrane-bound protein. Published values of K_d and n for each of the two sites vary, but it seems likely that the investigators concerned are studying the same proteins. Specificity of site I is poor; both active and inactive auxins compete for NAA binding. Site II, however, shows good specificity for active auxin analogues and antiauxins. By density gradient centrifugation and membrane marker location, site I has been assigned to the endoplasmic reticulum and site II to the plasma membrane or tonoplast. It has been suggested that the site located on the ER is a precursor of site II. Alternatively, the two sites may act in concert; lower selectivity of site I may occur simply because site II provides the primary molecular discrimination at the cell membrane. The distinction between the two sites has, however, been questioned. If the data are corrected for non-specific binding it has been argued that only a single class of binding sites can be observed with the kinetics described.

Sites I and II have been solubilized from their membranes by an acetone method and purified by FPLC and affinity chromatography. Auxin binding activity elutes from gel filtration columns at around 40–45 kD and has an isoelectric point in the range pI 4.5–5.2. Monospecific antibodies have been raised which recognize the affinity purified auxin binding protein. SDS gel electrophoresis of the immuneprecipitated auxin binding protein has revealed that the polypeptide has a molecular weight of 20–22 kD. These data suggest that the native protein is either a dimer, or a monomer which aggregates during gel filtration. The antibodies have also been used to immuneprecipitate a 20–22 kD polypeptide from the in-vitro translation products of coleoptile polyA$^+$ mRNA. This development should lead the way to cDNA cloning of the putative auxin receptor.

The cellular location of sites I and II has been investigated by immuno-light microscopy and binding assays of tissue extracts. There is good evidence that these binding sites exhibit tissue specificity. Immunologically the protein is located in maize coleoptile epidermal cells and is less abundant in leaves and stems than in the fast-growing tissues of root, tassel and silk. In addition, apical sections of oat coleoptiles are more auxin responsive than basal sections, and contain higher levels of binding activity. Perhaps the most convincing evidence for a receptor role of these binding sites is the observation that monospecific antibodies raised against them will inhibit auxin-induced growth.

It has been proposed that auxin-induced cell elongation is mediated via a stimulation in the activity of a proton pumping ATPase located within a cell membrane (see Chapter 10). In a series of intriguing experiments, Thompson *et al.* (1983) have demonstrated that the addition of ATP, the synthetic auxin NAA, and semi-purified receptor from maize tissue to reconstituted lipid membrane results in a significant increase in ion flux. In contrast, no stimulation is induced by the addition of the inactive auxin analogue benzoic acid (Figure 9.4). Suprisingly, these exciting results have yet to be followed up, even though they provide the first demonstration of a potential link between PGR binding to a putative receptor and the evocation of a biochemical response. Clearly, this study points to one direction which future work should take.

The best characterized soluble/nuclear-bound auxin binding site has been isolated from tobacco callus cultures. Cytosol of tobacco callus cells contains a specific high-affinity binding protein with a K_d for IAA of 1.6×10^{-8}M and in abundance of less than 200 fmol mg protein^{-1}. The binding site is also present in a non-covalently bound form in the nucleus. This protein can be reversibly phosphorylated, which alters its affinity for ligands and also provides a useful means of marking it with ^{32}P. The putative receptor has been purified by affinity chromatography of nuclear extracts. Gel filtration puts the molecular weight at approximately 120–125 kD. However, the receptor labelled with ^{32}P runs on polyacrylamide gels at 50 kD. The affinity-purified and phosphorylated binding protein has the capacity to stimulate RNA polymerase II activity in isolated nuclei providing auxin is also present in the medium. Furthermore, the stimulation of mRNA transcription is proportional to the degree of receptor occupancy. The significance of these observations is unclear, since subtler changes in gene expression are thought to be involved in auxin action (see Key *et al.*, 1986).

Although the signs that we are close to isolating a physiologically significant auxin receptor are optimistic, a number of molecules have the capacity to bind auxins. For instance, it has been demonstrated that

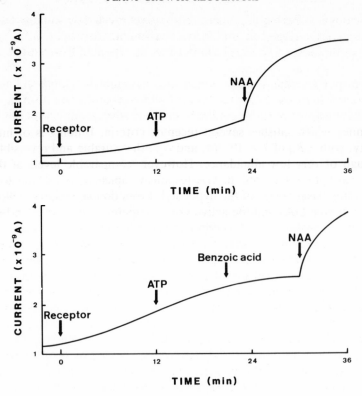

Figure 9.4 Electrochemical response of lipid membrane preparations to the addition of purified auxin receptor, ATP, and NAA (active auxin analogue) or benzoic acid (inactive auxin analogue) (Redrawn from Thompson *et al.* 1983.)

under certain assay conditions both ribulose-1,5-bisphosphate carboxylase (rubisco) and bovine serum albumen (BSA) can fulfil criteria (i)–(iv) outlined earlier for a putative receptor (for details see Libbenga *et al.*, 1985). Therefore, further efforts must be concentrated on coupling auxin binding to a cellular response to ensure that we are on the right track.

9.2.2 Gibberellins

Progress on the study of GA-binding proteins has been far less rewarding. This has been partly due to lack of general availability of high specific activity radiolabelled GAs and of pure preparations of

GA analogues. Perhaps for these reasons, relatively few workers have been actively engaged in this field of research. Nevertheless, in the last six years some advances have been made (for details see Stoddart, 1986).

Added GA stimulates the elongation of cucumber hypocotyls, and this growth response forms the basis of a bioassay for GAs. Cucumber hypocotyl cytosol contains a soluble protein which binds (^{3}H)-GA$_4$ in a manner which satisfies several receptor criteria. Binding is of high affinity, with a K_d of 7×10^{-8}M, and is both saturable and reversible, although of very low abundance. However, comparable levels of the protein are present in both the GA-responsive apical hook and the non-responding basal region of the hypocotyl. From double reciprocal plots it appears that GA$_4$ and the active GA$_7$ compete for the same binding site. Affinity ranking of GA derivatives and other PGRs has revealed an encouraging degree of structural specificity of the binding protein with respect to the biological activities of these ligands. However, the binding specificity is also consistent with that of substrates for the enzyme 2β-hydroxylase which converts GA$_1$ to GA$_8$. Therefore, it has been suggested that the binding protein may be a 2β-hydroxylase enzyme. Until antibodies to the 2β-hydroxylase enzyme become available it will not be possible to resolve this question.

It has been hypothesized that GAs might operate by directly interacting with membrane phospholipid. In support of this theory, differential scanning calorimetry and electron spin resonance techniques have detected membrane perturbations in liposomes after exposure to a GA$_4$/GA$_7$ mixture (Pauls et al., 1982). The response of the liposome membrane to GA$_8$, an inactive GA, was imperceptible. Although the concentrations of GA used in this study were very high, this intriguing result highlights the importance of keeping an open mind about the nature of receptor sites for PGRs.

9.2.3 Abscisic acid

ABA contains an α,β-unsaturated ketone group that is photoactivated by UV irradiation, probably to a highly reactive triplet excited state. This can abstract a hydrogen atom from a donor molecule, such as an amino acid and residue of a receptor protein, yielding two radicals which subsequently couple. In this way ABA may be covalently attached to its receptor protein. Hornberg and Weiler (1984) have used this technique for photoaffinity labelling of proteins in intact guard cell protoplasts of *Vicia faba* (Figure 9.5). In *Vicia* guard cells, the (+) enantiomer of ABA is active and the (−) enantiomer inactive. The latter compound is therefore a good and essential control in the affinity-labelling experiments. Encouragingly, there is a marked specificity for the active enantiomer by the protoplasts. Labelling is saturable, reversible with the active enantiomer, and of high affinity with a K_d of 3–4 nM. This corresponds closely to the 5 nM concentration required for half-maximal stomatal closure. The number of binding sites is estimated to be approximately 19.5×10^5 per protoplast. Since mild trypsin digestion of protoplasts eliminates binding, this suggests that the putative receptor proteins lie at the outer face of the plasmalemma. Both the kinetics and pH dependence of affinity binding are compatible with the physiological response of epidermal strips to ABA, and there is a close correlation between the biological activity of analogues of the PGR and their ability to displace (+)ABA from the binding sites. In addition, mesophyll protoplasts show an approximately 10-fold lower affinity for ABA, confirming some tissue specificity.

The ligand–receptor complex can be resolved by SDS polyacrylamide gel electrophoresis into three protein fractions having molecular weights of 20.2, 19.3 and 14.3 kD. The higher molecular weight fraction is preferentially labelled at high pH, presumably by $(+)ABA^-$, the others at acidic pH, presumably by protonated (+)ABA–H. This may explain why ABA can induce stomatal closure over a wide pH range. When the results of this study were published it was thought by many that the isolation and purification of an ABA receptor was imminent. Since then, however, further progress has been negligible.

9.2.4 Cytokinins

CKs occur both as free molecules in plant cells and also incorporated into tRNA (Figure 9.6). The more active *trans*-isomer is the predominant free form, while *cis*-zeatin is present in tRNA. The CK is located at the 3′ end of the anticodon sequence and influences the configuration of the tRNA molecule. It is generally accepted that the incorporated CK has a role in the interaction of the anticodon loop with ribosomal or mRNA binding sites and thus may influence mRNA translation. However, it is unlikely that such a mechanism can explain the observed physiological consequences of applying CKs to plant tissues. It is more likely that free CKs are active *per se*, and have a specific role in regulating plant development distinct from their function in tRNA.

Many researchers have investigated whether CKs interact with a re-

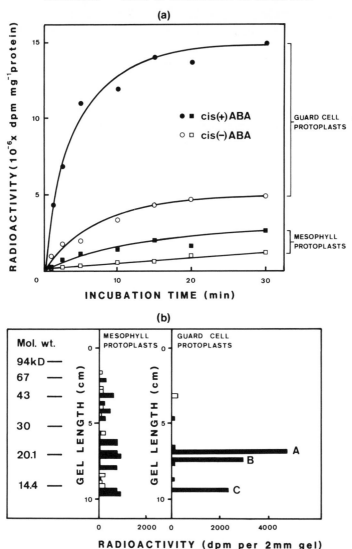

Figure 9.5 Photoaffinity labelling of ABA to leaf protoplasts. (*a*) Kinetics of binding of ^{3}H-*cis*(+) and ^{3}H-*cis*(−)ABA to guard cell or mesophyll protoplasts, (*b*) SDS-polyacrylamide gel electrophoresis of photoaffinity-labelled mesophyll or guard cell protoplast proteins after incubation with either ^{3}H-*cis*(+) ABA (solid bars) or ^{3}H-*cis*(−)ABA (open bars). Three proteins from guard cell protoplasts (designated *A, B* and *C*) cross-link specifically with ^{3}H-*cis*(+)ABA, and these proteins have molecular weights of 20.2, 19.3, and 14.3 kD respectively. (Data redrawn from Hornberg and Weiler, 1984.)

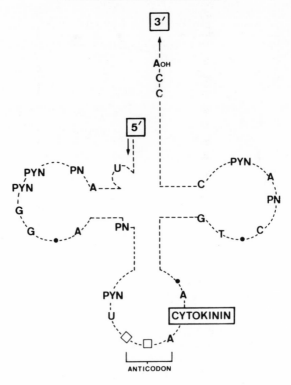

Figure 9.6 Location of cytokinin in tRNA molecule. A, adeneine; C, cytidine; G, guanine; U, uracil; T, thymidine; PN, purine nucleoside; PYN, pyrimidine nucleoside.

ceptor molecule as a prelude to exerting their physiological effects. However, no definitive candidate for a receptor has yet emerged. Low-affinity CK binding sites can be found on ribosomes with a K_d of 10^{-4}M. These sites are thought to reflect an affinity of ribosomal proteins for nucleotides. A high-affinity CK binding protein, K_d 6×10^{-7}M, has been found associated with wheat germ ribosomes. This has been designated CK binding factor 1 (CBF-1) and has been the subject of extensive study. It is an embryo-specific protein and has been purified by affinity chromatography on Sepharose 4B-benzyladenosine. It is a trimer composed of three identical 54 kD subunits with a single binding site per molecule of holoprotein. The binding residues have been located by photoaffinity labelling with (^{14}C)-2-azido-6-benzyladenine, partial proteolysis and separation by reverse-phase HPLC. Monospecific and monoclonal antibodies have been prepared to CBF-1, and an 800bp cDNA clone for the gene isolated (Fox and Brinegar, 1986). Unfortunately, however, the ontogeny of CBF-1 during seed development and germination, and its similarity in structure to vicilin, make it more likely that CBF-1 functions as a storage protein rather than as a CK receptor *per se*. Nevertheless, it might play a role in the sequestration of CK in the developing seed.

A number of other CK binding proteins have been reported but are less well characterized than CBF-1. These have been discussed by Venis (1985), and to date none is a convincing candidate as a receptor for CKs.

9.2.5 Ethylene

Ethylene binding sites have been discovered in several plant species, but in only two cases, cotyledons of French bean *Phaseolus vulgaris*, and mung bean *P. aureus*, have the proteins been characterized in detail (see Hall, 1986). Since the two systems have features in common, the discussion will be limited to the former species.

Cotyledons of *P. vulgaris* contain a protein which binds ethylene with a high affinity, K_d 1×10^{-10}M, although with low rate constants of association and dissociation. It increases in abundance from 3 to 40 pmol g tissue^{-1} as the cotyledons mature. Affinities of the protein for several structural analogues of ethylene are consistent with their relative physiological effectiveness. By cell fractionation and EM autoradiography of osmium fixed radiolabelled ethylene, the protein appears to be associated with the ER and protein bodies (Evans *et al.*, 1982). This membrane-bound hydrophobic protein has been purified and its molecular weight determined to be 50–60 kD. It is likely that this is an overestimate of its size since the protein is almost certainly complexed with the detergent used to solubilize it. At the present time we do not know if this protein is an ethylene receptor. In its favour are the high affinity and specificity for ethylene. Against it are, firstly, the fact that in this tissue there is no known physiological response to ethylene, and secondly the slow rates of dissociation and association. Classical receptor kinetics, and the fairly rapid switching of growth responses by ethylene, demand high rate constants of association and dissociation. Because ethylene is volatile, conventional binding studies will only be expected to reveal binding with low rates of association and dissociation. However, using antibodies raised against the identified ethylene binding protein,

F

it may be possible to probe for sites with more conventional receptor kinetics.

9.3 Sites of perception or deception?

Throughout this chapter it has been assumed that the primary action of a PGR is an interaction with a receptor protein, and that this constitutes the mechanism by which plant cells perceive their messages. In order to understand how PGRs regulate differentiation and development it is necessary to identify these receptor proteins. It will be clear to the reader that numerous plant proteins are capable of binding PGRs. It is against this background of often overwhelming nonspecific and quasi-specific binding that researchers are seeking to discover the few low-abundance proteins that possess genuine receptor characteristics. The difficulty of this task cannot be overemphasized. The fact that the PGR binding described in some instances clearly has no relevance to perception is an indication of the inherent difficulty of strictly applying the receptor criteria. At present, slow progress has been made in the search for receptors to GAs, CKs or ethylene. Nevertheless there are grounds for optimism as we have intimated in the previous section, and good reasons to believe that newly emerging techniques for dealing with low-abundance proteins will assist significantly in these studies. A good example of this is the way in which the search for ABA receptors has been transformed by the application of photoaffinity and protoplast techniques. Auxin receptors are clearly more amenable to the conventional binding assays, and there is every indication that the proteins described and presently being characterized may indeed be receptors. Absolute validation of this requires receptor criterion (vi) to be satisfied, i.e. that receptor stimulation is linked to a physiological response. Such responses are often complex, making this criterion difficult to address. In the next chapter we will consider those events which occur as a result of the perception of PGRs, and attempt to piece together their modes of action.

CHAPTER TEN

MECHANISMS OF ACTION — TOWARDS A MOLECULAR UNDERSTANDING

It is apparent from the previous chapters that PGRs have the capacity to elicit a variety of responses including cell division, differentiation, growth, and senescence in a wide range of organs, tissues and cells. The fundamental question that remains is by what mechanisms they act. For instance, does ABA induce subtle changes in the pattern of genes being expressed in the nuclei of developing soybean cells in the same way as it triggers the closure of stomata? Is the sequence of events the same when ethylene stimulates the degradation of the middle lamella during abscission and ripening? Do IAA and fusicoccin both promote H^+ efflux by the same mechanism? During the past few years, a number of important advances have been made in the elucidation of the mechanisms of PGR action, and it is evident that exciting progress in this area is imminent. In this chapter the current state of knowledge (and ignorance) of the molecular mechanism of action of PGRs is examined.

10.1 Regulation of ion movement

There is compelling evidence that PGRs can influence the properties of cell membranes and thus affect ion movement. Perhaps the best documented example of this is the stimulation of H^+ efflux in some tissues by auxins. The role of this phenomenon in the regulation of growth has been discussed previously (see Chapter 7). A number of researchers have tried to demonstrate that auxin stimulates H^+ efflux as a result of the PGR interacting with an ATPase. Attempts to simulate this using cell-free systems have met with mixed success, and have been greeted with justifiable reservations. However, a report by Scherer (1984) has convincingly demonstrated that both IAA and 2,4-D can stimulate the ATPase activity in plasma membrane-containing fractions from *Zucchini* hypocotyls at low ATP concentrations (Figure 10.1). The results from this study suggest that one of the effects of auxin is to decrease the apparent K_m of a plasmamembrane ATPase.

During water stress, ABA levels in leaves increase and stomata close. The rise in ABA is thought to be responsible for initiating stomatal

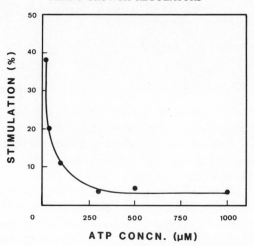

Figure 10.1 Dependence of ATPase stimulation by 10^{-6}M IAA on ATP concentration in a plasmamembrane-enriched fraction from pumpkin hypocotyls. (Replotted from Scherer, 1984.)

closure (see Chapter 8). Movements of K^+ between the guard cells and their neighbours play an important role in the production of the turgor changes necessary to regulate stomatal aperture. It has been shown that ABA causes a transient increase in the efflux of K^+ and Cl^- from guard cells of *Commelina communis* but has little impact on the influx of these ions. These effects of ABA on ion transport lower the osmotic potential of the cell and precipitate stomatal closure. ABA may influence the transport of other ions such as Ca^{2+} in the regulation of stomatal aperture, and these may act as 'second messengers' in the response (see section 10.3.1). Using *Vicia faba* guard-cell protoplasts, Schauf and Wilson (1987) have recently demonstrated that ABA significantly stimulates both the number of open K^+ channels and their duration of opening. These workers have probed the effects of ABA by 'patch-clamp analysis', and the use of this technique in other protoplast systems should dramatically advance our understanding of the effect of PGRs on ion transport over the next few years.

Ion transport can be influenced by a number of compounds other than PGRs. In particular, the fungal toxin fusicoccin (FC) has a dramatic effect. FC stimulates the H^+ efflux by plant cells more rapidly and to a greater extent than IAA. In addition it is a potent stimulator of elongation growth of a variety of tissues. The toxin also causes a marked stimulation of K^+ influx in isolated guard cells from *Commelina communis* leaves, and inhibits efflux of the ion (Figure 10.2). In keeping with these effects on ion movement, FC stimulates stomatal opening. It is apparent from the studies carried out on FC that the compound can affect the transport of many ions in the absence of an effect on gene expression (Marre, 1979). The capacity of FC to mimic PGRs is not restricted to growth and stomatal aperture. The compound can also promote cotyledon expansion like CKs, stimulate germination like GAs, and inhibit abscission and elevate ethylene biosynthesis like auxins. These observations raise the possibility that all these PGR effects might be regulated at the level of ion transport. Mystifyingly, the receptor for FC does not bind any of the known PGRs, and this raises the intriguing question of whether FC is a structural analogue of an important regulatory compound that has yet to be isolated.

10.2 Regulation of gene expression

The application of a number of sophisticated techniques for protein and nucleic acid identification and characterization has led to a rapid increase in our understanding of the effect of PGRs on gene expression. These techniques include:

(i)   *In-vivo* labelling of polypeptides, and *in-vitro* translation of mRNA followed by 1- or 2-dimensional gel electrophoresis, to determine PGR-induced changes in proteins synthesized and mRNA populations

(ii) cDNA cloning of poly(A^+) RNA and the selection of clones to specific PGR modulated RNA transcripts

(iii) cDNA–RNA hybridization to estimate levels of specific transcripts (this is a powerful technique to 'home in' on specific mRNA species amongst a background of many thousands of gene transcripts)

(iv) Cloning of genomic DNA and selection, using cDNAs, of genomic clones of genes that are regulated by PGRs

(v)   Sequencing of cDNA and genomic clones to provide nucleotide and derived amino acid sequences of genes of interest, and the nucleotide sequence of 5' (upstream) and 3' (downstream) non-coding regions that may contain regulatory sequence elements involved in responses to PGRs

(vi) Transformation of plant cells by the introduction of 'foreign' DNA.

In this section the current state of knowledge of the regulation of

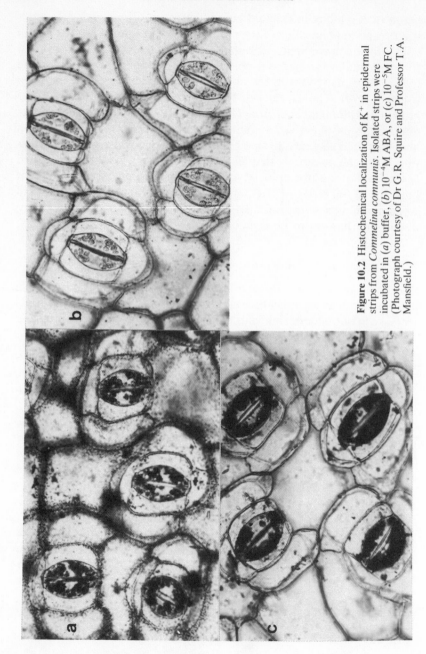

Figure 10.2 Histochemical localization of K$^+$ in epidermal strips from *Commelina communis*. Isolated strips were incubated in (*a*) buffer, (*b*) 10^{-4}M ABA, or (*c*) 10^{-5}M FC. (Photograph courtesy of Dr G. R. Squire and Professor T. A. Mansfield.)

gene expression by PGRs in a number of model systems will be considered. These systems have been referred to in previous chapters, but with a different emphasis.

10.2.1 Auxins

Subtle changes in the pattern of genes being expressed by soybean hypocotyls and pea stems during auxin-induced cell elongation have been described by several laboratories (see review by Theologis, 1986). cDNA hybridization studies have revealed that, of some 40 000 poly-(A^+)RNA species in the intact soybean hypocotyl, only a handful are up- or down-regulated by auxin. This effect of the PGR occurs within minutes of application. Two mRNAs present in the elongating region of the soybean hypocotyl appear to be genuinely up-regulated by auxin and these have been sequenced from genomic clones. They encode apparently related proteins of 20 and 30 kD. Although the identities and functions of these proteins are not known, nucleotide and amino acid sequence comparisons with documented genes could prove to be illuminating. Four other mRNAs up-regulated by 2,4-D in soybean hypocotyls have also been cloned. The results of *in-vitro* nuclear run on transcription experiments suggest that their expression can be attributed to increased gene transcription rather than post-transcriptional events. In fact, one of the clones increases in nuclei within 5 minutes of addition of auxin to the intact tissue. The remarkable speed of this response suggests, but does not prove, that it is closely linked to the primary action of auxins.

10.2.2 Gibberellins

Gibberellin induction of enzyme synthesis in the cereal aleurone layer (see Chapter 6) is one of the best known examples of the regulation of gene expression by a PGR. GAs induce *de novo* synthesis of α-amylase in aleurone layers, and ABA overcomes this effect. Approximately 70% of the protein synthesized in GA-induced aleurone cells is the product of the expression of approximately a dozen genes. In barley aleurone cells, levels of α-amylase mRNA increase within 1–2 h after applying GA_3, and become about 50 times more abundant than in control tissue (Figure 10.3). Nuclear run on experiments indicate that the increase in α-amylase mRNA appears to involve increased transcription of α-amylase genes (Figure 10.4). However, post-transcriptional effects of GA

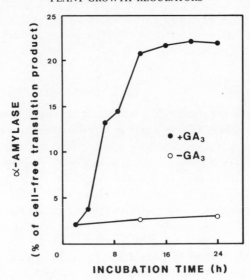

Figure 10.3 Time-course of α-amylase mRNA accumulation in barley aleurone cells after application of GA_3. (Redrawn from Higgins *et al.*, 1982.)

cannot be excluded, and indeed seem likely. Interestingly, the GA-induced accumulation of α-amylase mRNA is inhibited by cycloheximide. This observation implies that some essential protein is synthesized prior to the induction of amylase gene transcription.

The α-amylase genes are complex, consisting of two related gene families located on chromosome groups 6 and 7 in wheat, and 1 and 6 in barley. These genes have been sequenced from cDNA and genomic clones, and the relationship between them is discussed by Baulcombe (1987). The gene families encode two distinct groups of α-amylase iso-enzymes which differ in a number of respects, such as their sensitivity to EDTA, sulphydryl reagents and low pH, and their activity towards soluble and insoluble starch. There is a good degree of coordination in the expression of individual α-amylase gene family members as indicated by the relative levels of mRNA encoding each of the iso-zymes within groups. Furthermore, since the GA-regulated expression of other genes appears to be co-ordinated along with α-amylase, a common regulatory mechanism may exist. As yet, analysis of sequence data from regions upstream of a number of GA-regulated genes has not revealed a potential regulatory sequence. The controlling elements may therefore comprise non-contiguous nucleotides and be a feature of the

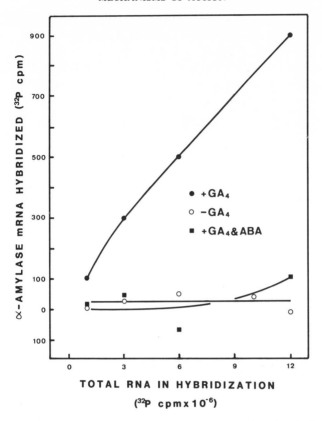

Figure 10.4 Hybridization of *in-vitro* synthesized RNA to α-amylase cDNA. Nuclei in the transcriptions were prepared from wild oat aleurone protoplasts which had been incubated for 72 h in 1 μM GA$_4$ or 1 μMGA$_4$ + 25 μM ABA. (Redrawn from Zwar and Hooley, 1985.)

structure adopted by DNA in chromatin. Transient expression experiments designed to identify regulatory elements are currently under way. There may be less strict co-ordination of the expression of amylase genes between the two families. During GA induction, both high and low pI mRNAs begin to increase at about the same time, but while the low pI mRNAs increase at a linear rate and are stimulated 20-fold, high pI mRNAs increase approximately 100-fold and then decline. In barley, but not in wheat, the two-gene families appear to respond differently to GA concentrations. Low pI mRNAs are enhanced by smaller concentrations of the PGR than high pI mRNAs, and are less dependent on changing concentrations of GA within the range 10^{-9} to 10^{-6} M. It should be emphasized that these studies are at the mRNA level and may not reflect gene transcription rates which could be identical for the different amylase genes. The effects may be related to mRNA stability and complicated by the presence of endogenous levels of the low pI mRNA. Nevertheless these differential effects may pose some intriguing questions about the mechanism of action of GA in controlling gene expression.

10.2.3 Abscisic acid

The GA-induced synthesis of α-amylase in the cereal aleurone can be inhibited by ABA. It is apparent that ABA prevents the GA-induced increase in transcription of α-amylase genes (Figure 10.4), although it is not clear whether this is a result of direct intervention in the action of GA, or by a less specific method. Interestingly, there is evidence that the action of ABA is dependent on transcription of other genes and may involve conversion of the PGR to phaseic acid (see Chapter 2). Other evidence of the effects of ABA on gene expression come from studies on seed development (see Chapter 6). Transcription rates of storage protein genes in *Brassica napus* embryos cultured *in vitro* are enhanced by treatment with ABA, and the implication is that this may occur *in vivo*. Similar *in-vitro* experiments with developing wheat embryos suggest that high levels of ABA present in the developing grain enhance the expression of genes encoding the lectin wheat germ agglutinin and other apparently globulin-related storage proteins, and an albumin-like protein found in mature embryos. These effects of ABA are thought to involve transcriptional and post-transcriptional events. In developing soybean cotyledons, ABA influences the level of transcript for the b-subunit of b-conglycin, although apparently it does not affect the levels of transcripts for the other two subunits of this storage protein.

10.2.4 Ethylene

Ethylene has been clearly demonstrated to regulate gene expression during ripening of both tomato and avocado fruit (see Chapter 8). A non-ripening tomato mutant, *rin*, does not synthesize the cell-wall degrading enzyme polygalacturonase (PG) (Figure 10.5), or accumulate a number of the ripening related mRNAs, even though the genes have been demonstrated to be present in the genomic complement of the plant. The *rin* mutant therefore must have a lesion in the chain of com-

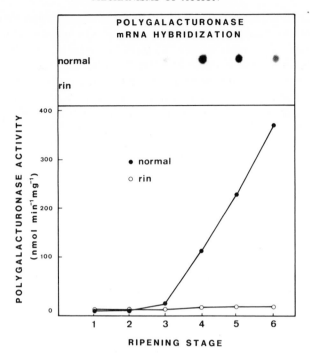

Figure 10.5 Polygalacturonase (PG) activity in normal and *rin* fruit at differing stages of ripening (1, mature green; 6, ripe). Upper portion of the figure shows the quantity of mRNA from the normal or mutant fruit that will hybridize to a radioactive cDNA probe for PG. (Data redrawn from Grierson *et al.*, 1986.)

mand leading to ethylene modulation of gene expression. If the ethylene binding sites described by Sisler (1982) are true receptors, then it would appear that *rin* is not a receptor mutant, since it contains the same concentration of these sites as normal plants. Intriguingly, PG is synthesized by *rin* plants during ethylene-induced leaf or flower abscission (Roberts *et al.*, 1987). The determination of this ripening lesion will undoubtedly take us a step closer to understanding the mechanism of action of ethylene in the regulation of gene expression during fruit ripening.

10.3 Second messengers

In the previous chapter it was argued that PGRs bind to specific receptors located in plant cells. But how does this event precipitate the action of a PGR? In the simplest system, the ligand–receptor complex may elicit a response directly. For instance, the receptor might be a membrane-bound protein responsible for the regulation of ion transport, which on interaction with the ligand undergoes a conformational change that initiates ion influx or efflux. Less directly, a ligand–receptor complex could migrate to the nucleus and interact with regions of DNA upstream from specific genes, thus altering their transcriptional rates. There is, however, reason to believe that some actions of PGRs are less direct than this, and involve the action of intermediaries called second messengers. A number of second messengers have been identified and characterized in animal cells. As these have been discovered, plant biologists have sought to invoke a role for them in the regulation of plant growth and development.

10.3.1 Calcium

There are two reasons why Ca^{2+} is considered to have useful signalling properties. Firstly, its concentration inside a cell is maintained 1000 to 10 000 times lower than that outside by specific membrane-bound pumps. Secondly, Ca^{2+} can reversibly modulate the activity of a number of specific Ca^{2+} binding proteins. The stimulation of mammalian cells by hormones can, by altering Ca^{2+} pumping, produce an increase in cytoplasmic Ca^{2+} and a concomitant activation of proteins involved in a biological response.

It is clear that Ca^{2+} also plays an important role in plant development (Hepler and Wayne, 1985), and may be involved in a number of events regulated by PGRs. For example, Ca^{2+} influences auxin-mediated growth responses by affecting both transport of the PGR and cell wall extensibility (see Chapters 4 and 7). However, neither of these responses represent the involvement of Ca^{2+} as a second messenger *per se*. One example of its involvement in PGR action is in the formation of buds on caulonema cells of the moss *Funaria* by CKs. Using the fluorescent Ca^{2+}-chelate probe, chlorotetracyclin, a specific increase in membrane-associated Ca^{2+} at the site of the CK-induced bud has been shown to precede the first bud cell division. In fact, bud initiation can be stimulated in the absence of CK by the manipulation of cellular Ca^{2+} levels with the aid of an ionophore. Alternatively, the process can be inhibited in the presence of the PGR by restricting Ca^{2+} availability. There is also evidence that the action of ABA on guard cells requires a free passage of Ca^{2+} into the cytosol, and it has been suggested that this may reflect the basic mechanism by which ABA acts.

10.3.2 Calmodulin

Calmodulin is a ubiquitous Ca^{2+} binding protein found in both plant and animal cells. On binding Ca^{2+}, it can activate a number of plant enzymes, including NAD-kinase and the Ca^{2+}-transporting ATPase. The demonstration that IAA-induced wheat coleoptile growth, GA_3-induced α-amylase synthesis in barley aleurones, ABA-promoted stomatal closure in *Commelina communis*, and CK-dependent betacyanin biosynthesis in *Amaranthus tricolor* cotyledons, can all be inhibited by calmodulin-binding drugs, is indirect evidence that the Ca^{2+}-calmodulin system could be involved in PGR action.

10.3.3 cAMP

In mammalian cells $3',5'$-cyclic AMP (cAMP) has been implicated as an important second messenger. In these systems, binding of a hormone to its receptor leads to the activation of adenylate cyclase and hence enhanced synthesis of cAMP. This cyclic nucleotide then activates cAMP-dependent protein kinases. These in turn modulate the activity of enzymes by phosphorylating them. Although cAMP has been identified in plant tissues, there is, as yet, no convincing evidence that cAMP is involved in the transduction of signals from PGRs.

10.3.4 Protein phosphorylation

In animal cells, the activity of key enzymes and receptors can be manipulated by phosphorylation and dephosphorylation (Cohen, 1985). A number of reports suggest that PGRs may activate proteins by phosphorylation mechanisms. Most convincingly, it has been demonstrated that phosphorylation and dephosphorylation modulate the affinity of the cytoplasmic auxin binding protein from *Nicotiana tabacum* to IAA (Van der Linde *et al.*, 1985). In addition, there have been reports that FC can stimulate the phosphorylation of a specific polypeptide in cultured sycamore cells, and the polyamine spermine can promote the phosphorylation of a 47 kD protein from isolated pea nuclei by 150%. However, the identity of these proteins remains unknown.

   Evidence is steadily accumulating that Ca^{2+} and Ca^{2+}-calmodulin protein kinases exist in plants (Poovaiah *et al.*, 1987). This may be the first sign that some of the pieces of the jigsaw are beginning to come together, although, as yet, no Ca^{2+}-dependent protein phosphatases have been identified in plant cells.

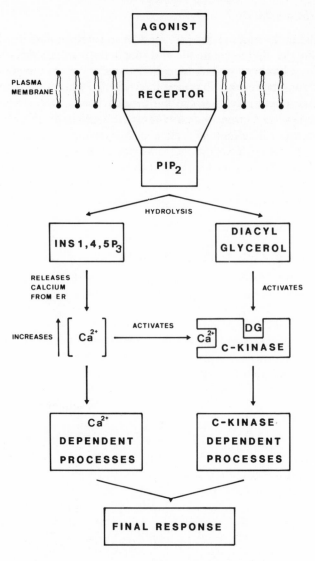

Figure 10.6 Proposed signal transduction pathway in animal systems. When an agonist binds to its receptor it leads to the hydrolysis of phosphatidylinositol 4,5-biphosphate (PIP$_2$) to give inositol 1,4,5-triphosphate (INS-1,4,5-P$_3$) and diacylglycerol (DG). The activity of these second messengers results in the mobilization of Ca^{2+} from intracellular stores such as the endoplasmic reticulum, and the activation of protein kinase C. The INS-1,4,5-P$_3$/Ca^{2+} and DG/C-kinase signal pathways are thought to co-operate with each other to control a spectrum of cellular responses and in some cells may act synergistically.

10.3.5 Inositol phospholipids

It has recently become established that inositol phospholipids are another link in the chain in the transduction of extracellular signals in animals. It has been demonstrated that, on binding to the plasma membrane, a hormone may stimulate the hydrolysis of phosphatidylinositol 4,5-bisphosphate, resulting in the production of diacylglycerol (DG) and inositol 1,4,5-triphosphate (IP$_3$). IP$_3$ then mobilizes Ca^{2+} from the endoplasmic reticulum, and DG activates a Ca^{2+}-dependent protein kinase (Berridge, 1987). With the recent demonstration that auxin can induce rapid changes in phosphatidylinositol metabolites in cells of *Catharanthus roseus* (Ettlinger and Lehle, 1988), the search is now on to determine whether PGRs could induce changes in cytoplasmic Ca^{2+} via IP$_3$-induced Ca^{2+} release (Figure 10.6).

10.4 Towards a molecular understanding

The past few years have witnessed a change of emphasis in PGR research that has been propelled by the emergence of new technologies for the study of proteins and nucleic acids. Today we can do more than just apply a PGR and quantify its effects. We can probe deeply into the mechanisms involved, and this can be done at various levels of PGR action. By identifying and characterizing receptors it is possible to determine the nature of PGR perception. Seeking out the mediators or second messengers will provide vital clues about PGR signal transduction. Characterizing PGR-responsive genes by sequence analysis and in functional assays of gene constructs in transient expression systems or transgenic plants may reveal the crucial *cis*-acting regulatory nucleotide sequences involved in their control. With the use of these sequences as affinity probes, it should be possible to isolate the *trans*-acting proteins involved in the regulation of gene transcription. In addition, the development of 'anti-sense mRNA' strategies may pave the way for both the identification of PGR responsive genes, and a method by which the control of their expression is regulated. The next few years should not only witness the elucidation of the molecular mechanism of a PGR, but should also be a tremendously exciting time to be working in this field of plant biology.

CHAPTER ELEVEN

COMMERCIAL APPLICATIONS OF PGRs — THOUGHT FOR FOOD?

A principal objective of the agricultural and horticultural industries is to manipulate plant growth and development so that the quantity or quality of a crop is increased. In the past, this has been achieved primarily through the skills of plant breeders, although the first use of chemicals for this purpose can be traced back to the mid-1930s. Since then there has been an increasing drive to identify chemicals with the ability to manipulate plant growth. Indeed it has been predicted that by the early 1990s the value of the commercial PGR market will have doubled from its present value to 1 000 million US dollars per annum at the producers' level, and will represent about 7% of the total worldwide sales of pesticides.

Advantages ascribed to the chemical manipulation of plant growth include the time-scale involved to achieve a commercially desirable result, the flexibility of the treatment, and the breadth of potential targets available. Furthermore, in contrast to a successful breeding programme, the 'beneficial' effects of a commercially viable chemical are usually not restricted to a single species. However, the initial euphoria that was associated with the concept of commercial PGRs has to some extent subsided under the pressure of the developmental costs of these chemicals and their impact on the natural environment. In addition, although a vast array of compounds has been screened for 'biological activity', the focus of the present range of commercial PGRs has become centred on growth retardation.

If plant differentiation and development is regulated *in vivo* by endogenous PGRs, then a knowledge of the mechanisms by which these compounds are synthesized, metabolized and act should assist in the design of chemical candidates for the manipulation of crop growth. In this chapter we examine how some commercial compounds act by mimicking or antagonizing endogenous PGRs, and consider in the light of our current state of knowledge the developmental processes that might be manipulated in the future.

11.1 Auxins and related compounds

One of the earliest uses of auxins for horticultural purposes was in the initiation and acceleration of the rooting of cuttings. The discovery of indole-3-acetic acid in the 1930s resulted in the development of the first group of commercial PGRs for this purpose. IAA became available in 1936, and this was rapidly followed by the synthetic auxins indolebutyric acid (IBA) and naphthylacetic acid (NAA). Both these compounds are more slowly metabolized than IAA by plant tissues, although the toxicity of NAA is high. Other auxin-like compounds have since been evaluated for their ability to enhance rooting; however, IBA has remained the compound of choice. Limited success has been achieved in the development of other practical applications for auxins. Synthetic auxins have been used to prevent fruit drop in apples and regulate the biennial bearing of fruit trees by their effects on thinning, but by far their chief use has been in their adoption as selective herbicides.

If plant tissues are exposed to supraoptimal concentrations of auxins, then root and shoot growth are inhibited, and cell differentiation and division are adversely affected. Prolonged exposure culminates in cell death. This is the underlying basis for the efficacy of the auxin herbicides which were developed in the 1940s. Auxin herbicides are selective in their action, killing only broad-leaved weeds within a cereal crop. The selectivity of the chemical seems to be primarily the result of elevated uptake by the broad-leaved species, although differential metabolism may be also partially responsible. The capacity of plants to metabolize butyric acid auxin derivatives has been used to advantage in the herbicide industry. Such butyric acid compounds are inactive auxins, but can be 'activated' by β-oxidation to acetic acid compounds. Plants lacking the β-oxidation pathway are resistant to butyric acid herbicides, whereas species having the pathway can be effectively induced to commit suicide by application of the herbicide.

The auxin transport inhibitor TIBA was first recognized as a potential regulatory chemical in 1942. However, it was not until 1961 that a detailed evaluation of its biological properties was carried out. TIBA has been found to increase lateral branch development in fruit trees and reduce plant height and stimulate branching in soybean. Although significant increases in yield of soybean have been found in field trials after TIBA application, primarily as a result of the reduction of crop lodging, extensive field evaluation has revealed that the effects of the chemical are unpredictable under different environmental conditions. Whilst

TIBA has not become adopted as a commercial PGR, it is apparent that its biological properties are consistent with a critical role for auxins in such developmental events as growth and apical dominance. Once the biosynthetic pathway of IAA has been irrefutably established, chemicals may be identified which have a specific inhibitory effect on auxin biosynthesis. The results obtained with TIBA suggest that such chemicals might prove to have potent biological properties, and may prove to have sufficiently reliable effects to become commercially viable PGRs.

11.2 Gibberellins and growth retardants

The majority of commercial formulations of GAs are restricted to use in specific horticultural situations. For instance, GAs are applied to increase yield and quality of table grapes, increase fruit set in apples and pears, and delay ripening of citrus fruits. Perhaps the most significant use is in the malting industry, where the PGR is routinely employed to stimulate the production of hydrolytic enzymes such as α-amylase to accelerate endosperm breakdown and the release of soluble sugars (see Chapter 6). GAs have been evaluated as potential commercial PGRs on a spectrum of crops, and although there is some evidence correlating application with an increase in yield, the results have not been consistent. In the absence of a reliable effect on a major crop it is unlikely that this PGR will be widely used as a commercial treatment, since both synthesis of the complex gibberellin skeleton and extraction of GAs from cell cultures of the fungus *Gibberella fujikuroi* are expensive.

The incorporation of dwarfing genes into modern wheats has had a dramatic effect on the agricultural industry. It has been proposed that stem shortening reduces lodging of the crop, thus increasing harvestable material, although additional factors may also contribute to the enhancement of yield (Quarrie and Gale, 1986). In other cereals, dwarfing genes have been unavailable to the breeder, and there has been an intense search for chemical retardants. A number of potent growth retardants have been identified, and in particular the compounds Chlormequat (2-chloroethyl-trimethylammonium chloride, CCC) and Paclobutrazol (PP333) have found favour as commercial PGRs. Both these compounds have been reported to inhibit the biosynthetic pathway of the GAs (Dalziel and Lawrence, 1984), which adds further support to the hypothesis that GAs are endogenous regulators of growth.

Since its discovery in the 1960s, CCC has become one of the most widely utilized commercial PGRs in the world. This is because the compound exhibits low toxicity and exerts its effects on many crop plants. It antagonizes GA biosynthesis by inhibiting the cyclization of geranyl-geranyl pyrophosphate into copalylphosphate (see Chapter 2), and can be absorbed either through the root or shoot tissues. PP333 was dis-covered more recently (1976), and is a member of the triazoles, which are the most highly active class of growth retardants documented so far (Figure 11.1). It is effective on a wide range of plant species, including several bulbous and woody plants, which have not proved sensitive to other retardants. The compound has been demonstrated to reduce the endogenous GA_1 level of wheat seedlings and block biosynthesis of the PGR in a cell-free system from *Cucurbita maxima* endosperm by inhibit-ing the activity of ent-kaurene oxidase (Hedden and Graebe, 1985).

Figure 11.1 Effect of $5\mu M$ paclobutrazol (PP333) on the growth of 6-day-old wheat (cv. Maris Huntsman) seedlings. Plants were untreated (left), or treated with the 2R,3R (middle) or the 2S,3S (right) enantiomer of PP333. The 2S,3S enantiomer inhibits stem elongation whereas the 2R,3R form has fungicidal activity but little impact on growth. Commercial PP333 is a 50:50 mixture of the 2R,3R and the 2S,3S enantiomers. (Photograph courtesy of Dr John Lenton.)

PP333 is xylem-mobile, and therefore effective retardation relies primarily on uptake of the chemical via the roots. Unfortunately the effects of PP333 can persist within the soil, which may limit its suitability as a growth retardant in some situations.

Although the most consistent effect of CCC and PP333 is on growth retardation (Figure 11.1), there have been reports that these chemicals can induce additional effects such as an increase in ear or grain number or modification in canopy structure, or confer limited fungicidal protection. Indeed, other members of the triazole family have been commercially developed into fungicides, such as 'Bayleton'. The fungicidal properties of the triazoles are due to their capacity to inhibit the biosynthesis of ergosterol, which is a vital component of fungal membranes. The biosynthetic pathway of sterols and GAs have features in common, and both CCC and PP333 have been reported to reduce the sterol levels of plant tissues (Lurssen, 1987). The functional role of sterols in membranes and the regulatory properties of the brassinosteroids may contribute to the spectrum of effects that these growth retardants can exert under certain conditions.

While the commercial use of CCC has been mainly restricted to cereals, PP333 is being extensively used in the horticultural industry to regulate the growth of fruit trees and ornamentals. In addition, there is considerable interest in using the latter to regulate the growth of grass and trees in amenity areas. The 'greens' market is of considerable commercial significance since it includes such sites as grass verges, golf-course fairways, cemeteries, and other areas difficult to mow.

The search for plant growth retardants is perhaps the major success story of the commercial PGR industry to date. It is clearly more than a coincidence that two of the most effective retardants have proved to be associated with their ability to inhibit GA biosynthesis; this highlights the significance of this PGR in the regulation of plant growth. These chemicals were discovered by screening procedures; however, as our knowledge of the enzymes involved in the biosynthetic pathway of GAs improves, it may be possible to predict the chemical structure of growth inhibitors 'tailored' to home in specifically on key enzymes of the GA biosynthesis pathway, and to suit particular species and environments.

11.3 Cytokinins and related compounds

Synthetic CKs such as kinetin and benzyladenine have been extensively examined for potential commercial exploitation. The ability of these compounds to promote lateral bud growth (see Chapter 7) has been utilized by the horticultural industry as an alternative to removal of the apex for the enhancement of flower production in roses and carnations. In addition, some success has been reported in the prevention of vegetable senescence by the application of CKs. However, apart from these examples and a few other specialized uses, CKs have not been exploited by the agrochemical industry. This is perhaps a reflection of both the cost of synthesis of CK-like compounds and the inability to locate an appropriate market that would benefit from their biological activity. As Garrod points out (1982), a more fruitful approach, based on experience with gibberellins, might be to screen chemicals for anti-CK activity. Certainly there is a suggestion that compounds such as this may prove to be effective in promoting stomatal closure and hence act as anti-transpirants (see section 11.4).

11.4 Abscisic acid and anti-transpirants

There is convincing evidence that stomatal aperture may influence such phenomena as water consumption, chilling injury, pollutant uptake and senescence of plants. The ability to manipulate stomatal movements, could therefore, have a significant effect on plant growth and development (see Fenton et al., 1982). Since ABA has been implicated in the regulation of stomatal closure (see Chapter 7), effort has been concentrated on determining the commercial potential of this endogenous PGR particularly with regards to its efficacy as an anti-transpirant.

Although ABA has been shown to have 'adequate' anti-transpirant properties on crops such as barley and coffee, its effects are short-lived (approximately 10 days) as a result of rapid metabolism and photoisomerization of the PGR. Frequent applications would therefore be necessary to protect plants from water stress, and this would be both inconvenient and expensive. Structural analogues of ABA have been synthesized but found to have similar limitations. Other chemicals have been screened for anti-transpirant activities and of these, the compound farnesol has proved to be one of the most effective. This compound was originally isolated from water-stressed *Sorghum* plants, and its accumulation was closely correlated with stomatal closure. Unfortunately, application of farnesol at concentrations sufficient to induce stomatal closure, can induce damage to the leaf mesophyll tissue which has restricted its commercial utilization. The mechanism by which farnesol stimulates stomatal closure is unknown, although it has been shown to elevate the level of ABA in spinach leaf tissues.

The conclusion from these studies is that, although there maybe considerable scope in having the ability to regulate stomatal movement at will, as yet our knowledge does not give us that privilege. An alternative approach to the elevation of ABA to precipitate stomatal closure may be to antagonize the effects of CKs (see Chapter 7). Indeed, some anti-cytokinins are currently being evaluated for anti-transpirant properties.

A number of reports have been published highlighting a positive relationship between endogenous or exogenous ABA and the accumulation of photosynthetic assimilates by seed tissues. These observations have been restricted to laboratory-grown plants and have not been substantiated in rigorous field tests. It may be that, under field conditions, ABA levels do not limit the assimilation of photosynthates by seeds. However, these effects of ABA are of potential significance, since demonstrations of a direct effect of PGRs on crop yield are rare. As our understanding of the role of ABA in seed maturation becomes clearer (see Chapter 6), it may be possible to use this information to develop a chemical treatment to elevate the efficiency of assimilate uptake by seed tissues in the field.

11.5 Ethylene-generating or suppressing compounds

The use of smoke from burning vegetation to stimulate pineapple flowering in the last century is probably one of the earliest documented examples of the use of a chemical to manipulate crop growth. Whilst the 'active' constituent in the smoke was identified as ethylene as early as the 1920s, the gaseous nature of this endogenous PGR limited its commercial development until an ethylene-releasing chemical was identified some 40 years later. The compound, ethephon (2-chlorethylphosphonic acid), is chemically stable below pH 4.1, but, on entering plant cells which are less acidic, liberates ethylene into the tissues. Ethephon is now one of the most widely used commercial PGRs (see Morgan, 1986) and it has been employed to regulate developmental events such as ripening, senescence and abscission in a range of agricultural and horticultural species. Moreover, it has specific applications, including the promotion of flower initiation in bromeliads, the modification of sex expression in cucumbers, the stimulation of latex flow in rubber trees, and most recently as 'Cerone' to inhibit growth in cereals. The variety of uses that have been adopted for ethephon may be a reflection of the spectrum of developmental processes which can be influenced by ethylene.

The phenomenon of tissue sensitivity to endogenous PGRs (see Chapter 4), has been used to advantage in the application of ethephon to cotton. This crop suffers from the requirement for multiple harvesting, and in addition, the utilization of late season flower buds and young fruit as sustenance for over-wintering insect pests. Since young fruit and flower buds are more sensitive to ethylene-induced abscission than mature fruit or leaves, prior to harvest the crop can be treated with ethephon without inducing defoliation. This treatment, which removes the young fruit and flower buds, also promotes uniform boll dehiscence and thereby allows once-over harvesting techniques to be employed. In this way the cotton crop can be subjected to a managerial control which helps to minimize losses.

The identification of ACC as a biosynthetic precursor of ethylene (see Chapter 2) has resulted in the development of a new class of ethylene-releasing chemicals. Interest has been centred on the generation of ACC derivatives which have modified capacities of uptake, transport, metabolism or ethylene release, since these parameters ultimately dictate the efficacy of a commercial ethylene-releasing PGR under different field conditions. Although a number of derivatives have been synthesized which exhibit biological properties, as yet detailed evaluations of their performances in the field have proved disappointing (Lurssen and Konze, 1985).

It is evident that the utilization of ethylene-releasing chemicals by the agricultural and horticultural industries has proved to be a highly successful ploy for the manipulation of plant development. Since ethylene appears to play a strategic role in regulating processes such as ripening, senescence, and abscission (see Chapter 8), there could be commercial potential in reducing levels of this PGR or suppressing its mode of action. The most potent inhibitors of ethylene biosynthesis are the compounds rhizobitoxin and AVG which both act by suppressing the activity of ACC synthase (see Chapter 2). However, the cost of these has so far precluded a detailed study of their impact on crop development. Although the idea of chemically reducing ethylene levels is appealing, the commercial development of this scheme has yet to be implemented. Clearly there is scope for further research in this area, since significant financial resources are currently employed in building 'controlled atmosphere' facilities for storage and transport of fruit and vegetables. These stores use CO_2 as a competitive inhibitor of ethylene, and also use scrubbers to minimize the level of ethylene in the atmosphere surrounding the produce. As a result the rate of ripening or senescence of the produce is reduced.

Figure 11.2 Effect of silver thiosulphate (STS) on carnation flowers. Blooms were maintained in a solution of sodium thiosulphate (control) or STS for 10 days.

Apart from CO_2, other compounds have been examined for their capacity to antagonize ethylene action. The most effective inhibitor of ethylene action so far documented is the silver ion, which is thought to prevent ethylene from successfully binding to its receptor. Silver ions are virtually immobile in plant tissues, and therefore commercial use has been made of the mobile silver thiosulphate complex (STS). Postharvest treatment of carnations by dipping the stems in STS has become a widely employed treatment in the horticultural industry to enhance the life of cut flowers (Figure 11.2), and this allows their long-term storage (see Nichols and Frost, 1985). In addition, STS is routinely sprayed on potted plants prior to shipment to minimize flower bud and petal abscission and epinastic bending of the leaves.

It has been convincingly demonstrated that, by elevating or suppressing ethylene levels or action, plant development may be manipulated to advantage. Although ethylene-releasing chemicals are commercially available, no effective compounds have been marketed to inhibit endo-genous production of this PGR. Further studies on the biosynthetic pathway of the gas may prove fruitful in this respect particularly with regard to the antagonism of ACC synthase activity. In addition, knowl-edge of the mechanism of ethylene perception by plant cells should aid attempts to regulate ethylene action. It is clear that, as further ethylene-related chemicals are developed, these will have a highly significant impact on the agricultural and horticultural industries.

11.6 Thought for food — food for thought

In theory, the chemical manipulation of plant growth and development is an approach which has considerable potential for the quantitative or qualitative improvement of crop performance. In practice, however, this method has proved difficult to exploit successfully. Perhaps this is not too surprising, since the chemicals that have been used have primarily been selected for on the basis of their biological activity in laboratory screens, and the mechanism by which they act has attracted only cursory attention. As a result, problems have been encountered over their specificity, reliability, toxicity and the persistence of chemical residues.

In this book we have sought to examine critically the role of specific endogenous chemicals or PGRs in plant growth and development. It is clear that whilst some stories are more complete than others, there is evidence in favour of a strategic role for PGRs in the regulation of a variety of developmental processes. Furthermore, an examination of the most successful commercial PGRs reveals that the biological pro-perties of these chemicals rely firmly on their ability to mimic or anta-gonize PGR activity. Therefore, as our knowledge of PGR biosynthesis and mode of action improves, it should be possible to exploit this infor-mation to identify chemicals with greater biological potency, specificity and reliability. However, the use of chemicals to manipulate plant de-velopment has inherent difficulties. For instance, it relies on the ability to deliver the compounds to a precise site of action. Considerable re-search effort is needed therefore to gain some control over both inter- and intracellular targeting of PGRs.

The current emphasis on plant molecular biology, will, in the future, lead to the opportunity of manipulating crop growth using the tech-niques of genetic engineering. This has certain advantages over chemi-cal regulation in that targeting is unnecessary and the direct stimulation of yield more feasible. However, use of transgenic plants does not allow the grower flexibility under conditions of environmental fluctuation. One solution to this problem would be to use chemicals to manipulate gene expression in transgenic plants. Such chemicals would have to have the capacity to interact directly or indirectly with the genome *in vivo*, so that they could regulate the expression of the tailored gene(s). Our knowledge of the action of PGRs suggests that these endogenous chemicals could be ideally suited for the job.

REFERENCES AND FURTHER READING

Chapter 2

Adams, D.O. and Yang, S.F. (1979) Ethylene biosynthesis: identification of 1-amino-cyclopropane-1-carboxylic acid as an intermediate in the conversion of methionine to ethylene. *Proc. Natl Acad. Sci. USA* **76**, 170–174.

Bandurski, R.S. (1984) Metabolism of indole-3-acetic acid. In *The Biosynthesis and Metabolism of Plant Hormones*, SEB Seminar Series 23, eds. A. Crozier and J.R. Hillman, Cambridge University Press, Cambridge, 183–200.

Beyer, E.M. (1985) Ethylene metabolism. In *Ethylene and Plant Development*, eds. J.A. Roberts and G.A. Tucker, Butterworth, London, 125–137.

Bleecker, A.B., Kenyon, W.H., Somerville, S.C. and Kende, H. (1986) Use of mono-clonal antibodies in the purification and characterization of 1-aminocyclopropane-i-carboxylate synthase, an enzyme in ethylene biosynthesis. *Proc. Natl Acad. Sci. USA* **83**, 7755–7759.

Burch, L.R. and Stuchbury, T. (1987) Activity and distribution of enzymes that intercon-vert purine bases, ribosides and ribotides in the tomato plant and possible implications for cytokinin metabolism. *Physiol. Plant.* **69**, 283–288.

Cohen, J.D. and Bialek, K. (1984) The biosynthesis of indole-3-acetic acid in higher plants. In *The Biosynthesis and Metabolism of Plant Hormones*, SEB Seminar Series 23, eds. A. Crozier and J.R. Hillman, Cambridge University Press, Cambridge, 165–181.

Coolbaugh, R.C. (1983) Early stages of gibberellin biosynthesis. In *The Biochemistry and Physiology of Gibberellins*, Vol. 1, ed. A. Crozier, Praeger, New York, 53–98.

Creelman, R.A. and Zeevart, J.A.D. (1984) Incorporation of oxygen into abscisic acid and phaseic acid from molecular oxygen. *Plant Physiol.* **75**, 166–169.

Graebe, J.E. (1987) Gibberellin biosynthesis and control. *Ann. Rev. Plant Physiol.* **38**, 419–465.

Horgan, R. (1984) Cytokinins. In *Advanced Plant Physiology*, ed. M.B. Wilkins, Pitman UK, London, 53–75.

Linforth, R.S.T., Bowman, W.R., Griffin, D.A., Marples, B.A. and Taylor, I.B. (1987) 2-*trans* ABA alcohol accumulation in the wilty tomato mutants *flacca* and *sitiens*. *Plant Cell Environm.* **10**, 599–606.

Loveys, B.R. and Milborrow, B.V. (1984) Metabolism of abscisic acid. In *The Biosyn-thesis and Metabolism of Plant Hormones*, SEB Seminar Series 23, eds. A. Crozier and J.R. Hillman, Cambridge University Press, Cambridge, 71–104.

Manning, K. (1986) Ethylene production and β-cyanoalanine synthase activity in carna-tion flowers. *Planta, Berlin* **168**, 61–66.

McGaw, B.A., Scott, I.M. and Horgan, R. (1984) Cytokinin biosynthesis and meta-bolism. In *The Biosynthesis and Metabolism of Plant Hormones*, SEB Seminar Series 23, eds. A. Crozier and J.R. Hillman, Cambridge University Press, Cambridge, 105–133.

Miyazaki, J.H. and Yang, S.F. (1987) The methionine salvage pathway in relation to ethylene and polyamine biosynthesis. *Physiol. Plant.* **69**, 366–370.

Phinney, B.O. (1984) Gibberellin A_1, dwarfism and the control of shoot elongation in higher plants. In *The Biosynthesis and Metabolism of Plant Hormones*, SEB Seminar Series 23, eds. A. Crozier and J.R. Hillman, Cambridge University Press, Cambridge, 17–41.

Phinney, B.O., Freeling, M., Robertson, D.S., Spray, C.R. and Silverthorne, J. (1986)

Dwarf mutants in maize — the gibberellin biosynthetic pathway and its molecular future. In *Plant Growth Substances 1985*, ed. M. Bopp, Springer, Berlin, 55–73

Sanders, I.O., Smith, A.R. and Hall, M.A. (1986) Ethylene metabolism and action. *Physiol. Plant.* **66**, 723–726.

Smith, A.R. and Hall, M.A. (1984) Biosynthesis and metabolism of ethylene. In *The Biosynthesis and Metabolism of Plant Hormones*, SEB Seminar Series 23, eds. A. Crozier and J.R. Hillman, Cambridge University Press, Cambridge, 201–229.

Smith, T.A. (1985) Polyamines. *Ann. Rev. Plant Physiol.* **36**, 117–143.

Sponsel, V.M. (1985) Gibberellins in *Pisum sativum* — their nature, distribution and involvement in growth and development of the plant. *Physiol. Plant.* **65**, 533–538.

Taylor, I.B. (1987) ABA-deficient mutants. In *Developmental Mutants in Higher Plants*, SEB Seminar Series 32, eds. H. Thomas and D. Grierson, Cambridge University Press, Cambridge, 197–217.

Thomashow, M.S., Hugly, W., Buchholz, and Thomashow, L. (1986) Molecular basis for the auxin-independent phenotype of crown gall tumor tissues. *Science*, **231**, 616–618.

Weiler, E.W. and Schroder, J. (1987) Hormone genes and crown gall disease. *Trends Biochem. Sci.* **12**, 271–275.

Further reading

Crozier, A. and Hillman, J.R. (eds.) (1984) *The Biosynthesis and Metabolism of Plant Hormones*, SEB Seminar Series 23, Cambridge University Press, Cambridge.

Chapter 3

Bassi, P.K. and Spencer, M.S. (1985) Comparative evaluation of photoionization and flame ionization detectors for ethylene. *Plant Cell Environm.* **8**, 161–165.

Davis, G.C., Hein, M.B., Chapman, D.A., Neely, B.C., Sharp, C.R., Durley, R.C., Biest, D.K., Heyde, B.R. and Carnes, M.G. (1986) Immunoaffinity columns for the isolation and analysis of plant hormones. In *Plant Growth Substances 1985*, ed. M. Bopp, Springer, Berlin, 44–51.

Hedden, P. (1986) The use of combined gas chromatography–mass spectrometry in the analysis of plant growth substances. In *Modern Methods of Plant Analysis*, Vol. 3, *Gas Chromatography/Mass Spectrometry*, eds. H.F. Linskens and J.F. Jackson, Springer, Berlin, 1–22.

Knox, J.P., Beale, M.H., Butcher, G.W. and MacMillan, J. (1987) Preparation and characterization of monoclonal antibodies which recognize different gibberellin epitopes. *Planta (Berlin)*, **170**, 86–91.

Mertens, R., Deus-Neumann, R. and Weiler, E.W. (1983) Monoclonal antibodies for the detection and quantitation of the endogenous plant growth regulator, ABA. *FEBS Lett.* **160**, 269–272.

Perrot-Rechenmann, C. and Gadal, P. (1986) Enzyme immunohistochemistry. In *Immunology in Plant Science*, SEB Seminar Series 29, ed. T.L. Wang, Cambridge University Press, Cambridge, 59–88.

Reeve, D.R. and Crozier, A. (1980) Quantitative analysis of plant hormones. In *Encyclopedia of Plant Physiology, New Series*, Vol. 9, *Hormonal Regulation of Development I*, ed. J. MacMillan, Springer, Berlin, 203–280.

Robins, R.J. (1986) The measurement of low-molecular-weight, non-immunogenic compounds by immunoassay. In *Modern Methods of Plant Analysis*, Vol. 4, *Immunology in Plant Science*, eds. H.F. Linskens and J.F. Jackson, Springer, Berlin, 86–141.

Ward, T.M., Wright, M., Roberts, J.A., Self, R. and Osborne, D.J. (1978) Analytical procedures for the assay and identification of ethylene. In *Isolation of Plant Growth Substances*, SEB Seminar Series 4, ed. J.R. Hillman, Cambridge University Press, Cambridge, 135–151.

Weiler, E.W., Eberle, J., Mertens, R., Atzorn, R., Feyerabend, M., Jourdan, P.S., Arnscheidt, A. and Wieczorek, U. (1986) Antisera and monoclonal antibody-based immunoassay of plant hormones. In *Immunology in Plant Science*, SEB Seminar Series 29, ed. T.L. Wang, Cambridge University Press, Cambridge, 27–58.

Wright, M. and Doherty, P. (1985) Auxin levels in single half nodes of *Avena fatua* estimated using high performance liquid chromatography with coulometric detection. *J. Plant Growth Reg.* **4**, 91–100.

Yokota, T., Murofushi, N. and Takahashi, N. (1980) Extraction, purification and identification. In *Encyclopedia of Plant Physiology, New Series*, Vol. 9, *Hormonal Regulation of Development I*, ed. J. MacMillan, Springer, Heidelberg and Berlin, 113–201.

Further reading

Crozier, A., Sandberg, G., Monteiro, A.M. and Sundberg, B. (1986) The use of immunological techniques in plant hormone analysis. In *Plant Growth Substances 1985*, ed. M. Bopp, Springer, Berlin, 13–21.

Davis, G.C., Hein, M.B., Neely, B.C., Sharp, C.R. and Carnes, M.G. (1985) Strategies for the determination of plant hormones. *Anal. Chem.* **57**, 638A–648A.

Ernst, D. (1986) Radioimmunoassay and gas chromatography/mass spectrometry for cytokinin determination. In *Modern Methods of Plant Analysis*, Vol. 4, *Immunology in Plant Sciences*, eds. H.F. Linskens and J.F. Jackson, Springer, Berlin, 18–49.

Rivier, L. and Crozier, A. (1987) *Principles and Practice of Plant Hormone Analysis*, Vol. 1 and 2.

Chapter 4

Boller, T. and Kende, H. (1980) Regulation of wound ethylene synthesis in plants. *Nature* (*London*), **286**, 259–260.

Bradford, K. J. and Yang, S.F. (1980) Xylem transport of 1-amino-cyclopropane-1-carboxylic acid, an ethylene precursor, in waterlogged tomato plants. *Plant Physiol.* **65**, 322–326.

Davies, W.J., Metcalfe, J.C., Schurr, U., Taylor, G. and Zhang, J. (1987) Hormones as chemical signals involved in root to shoot communication of effects of changes in the soil environment. In *Hormone Action in Plant Development — A Critical Appraisal*, eds. G.V. Hoad, J.R. Lenton, M.B. Jackson and R.K. Atkin, Butterworth, London, 201–216.

Firn, R.D. (1986) Growth substance sensitivity: The need for clearer ideas, precise terms and purposeful experiments. *Physiol. Plant.* **67**, 267–272.

Goldsmith, M.H.M. (1977) The polar transport of auxin. *Ann. Rev. Plant Physiol.* **28**, 439–478.

Hartung, W., Heilmann, B. and Gimmler, H. (1981) Do chloroplasts play a role in abscisic acid synthesis? *Plant Sci. Lett.* **22**, 235–242.

Hertel, R., Lomax, T.L. and Briggs, W.R. (1983) Auxin transport in membrane vesicles from *Cucurbita pepo* L. *Planta* (*Berlin*), **157**, 193–201.

Jacobs, M. and Gilbert, S.F. (1983) Basal localization of the presumptive auxin transport carrier in pea stem cells. *Science*, **220**, 1297–1300.

Jacobs, M. and Short, T.W. (1986) Further characterization of the presumptive auxin transport carrier using monoclonal antibodies. In *Plant Growth Substances 1985*, ed. M. Bopp, Springer, Berlin, 218–226.

Milborrow, B.V. and Rubery, P.H. (1985) The specificity of the carrier-mediated uptake of ABA by root segments of *Phaseolus coccineus* L. *J. exp. Bot.* **36**, 807–822.

Nissen, P. (1985) Dose response of auxins. *Physiol. Plant.* **65**, 357–374.

Osborne, D.J. (1982) The ethylene regulation of cell growth in specific target tissues of plants. In *Plant Growth Substances 1982*, ed. P.F. Wareing, Academic Press, London and New York, 279–290.

Osborne, D.J., McManus, M.T. and Webb, J. (1985) Target cells for ethylene action. In *Ethylene and Plant Development*, eds. J.A. Roberts and G.A. Tucker, Butterworth, London, 197–212.

Riov, J. and Goren, R. (1979) Effect of ethylene on auxin transport and metabolism in midrib sections in relation to leaf abscission of woody plants. *Plant Cell Environm.* **2**, 83–89.

Trewavas, A.J. (1982) Growth substance sensitivity: the limiting factor in plant development. *Physiol. Plant.* **55**, 60–72.

Turnbull, C.G.N. and Hanke, D.E. (1985) The control of bud dormancy in potato tubers. Evidence for the primary role of cytokinins and a seasonal pattern of changing sensitivity to cytokinin. *Planta (Berlin)*, **165**, 359–365.

Weiler, E.W., Schnabl, H. and Hornberg, C. (1982) Stress-related levels of abscisic acid in guard cell protoplasts of *Vicia faba* L. *Planta (Berlin)*, **154**, 24–28.

Wright, M. (1981) Reversal of the polarity of IAA transport in the leaf sheath base of *Echinochloa colonum. J. exp. Bot.* **32**, 159–169.

Yoshii, H. and Imaseki, H. (1982) Regulation of auxin-induced ethylene biosynthesis. Repression of inductive formation of ACC-synthase by ethylene. *Plant Cell Physiol.* **23**, 639–649.

Further reading

Rubery, P.H. (1987) Modulation of hormone transport in physiological studies. In *Hormone Action in Plant Development — a Critical Appraisal*, eds. G.V. Hoad, J.R. Lenton, M.B. Jackson and R.K. Atkin, Butterworth, London, 161–174.

Trewavas, A.J. (1987) Sensitivity and sensory adaptation in growth substance responses. In *Hormone Action in Plant Development — a Critical Appraisal*, eds. G.V. Hoad, J.R. Lenton, M.B. Jackson and R.K. Atkin, Butterworth, London, 19–38.

Chapter 5

Aloni R. (1987) Differentiation of vascular tissues. *Ann. Rev. Plant Physiol.* **38**, 179–204.

Evans, L.T. (1987) Towards a better understanding and use of the physiology of flowering. In *Manipulation of Flowering*, ed. J.G. Atherton, Butterworth, London, 409–423.

Francis, D. (1987) Effects of light on cell division in the shoot meristem during floral evocation. In *Manipulation of Flowering*, ed. J.G. Atherton, Butterworth, London, 289–300.

Holder, N. (1979) Positional information and pattern formation in plant morphogenesis and a mechanism for the involvement of plant hormones. *J. theoret. Biol.* **77**, 195–212.

Jaffe, M.J., Bridle, K.A. and Kopcewicz, J. (1987) A new strategy for the identification of native plant photoperiodically regulated flowering substances. In *Manipulation of Flowering*, ed. J.G. Atherton, Butterworth, London, 279–287.

Jacobs, W.P. (1984) Function of hormones at tissue level of organization. In *Encyclopedia of Plant Physiology, New Series*, Vol. 10, *Hormonal Regulation of Development II*, ed. T.K. Scott, Springer, Berlin, 149–171.

Pearce, D., Miller, A.R., Roberts, L.W. and Pharis, R.P. (1987) Gibberellin-mediated synergism of xylogenesis in lettuce pith cultures. *Plant Physiol.* **84**, 1121–1125.

Pharis, R.P. and King, R.W. (1985) Gibberellins and reproductive development in seed plants. *Ann. Rev. Plant Physiol.* **36**, 517–568.

Sachs, T. (1986) Cellular interactions in tissue and organ development. In *Plasticity in Plants*, SEB Symposium Vol. 40, eds. D.H. Jennings and A.J. Trewavas, Company of Biologists, Cambridge, 181–210.

Schwabe, W.W. (1987) Hormone involvement in daylength and vernalization control of reproductive development. In *Hormone Action in Plant Development — a Critical Appraisal*, eds. G.V. Hoad, J.R. Lenton, M.B. Jackson and R.K. Atkin, Butterworth, London, 217–230.

Thomas, B. and Vince-Prue, D. (1984) Juvenility, photoperiodism and vernalisation. In *Advanced Plant Physiology*, ed. M.B. Wilkins, Pitman UK, London, 408–439.

Tran Thanh Van, K., Toubart, P., Cousson, A., Darvill, A.G., Collin, D.J., Chelf, P. and Albersheim, P. (1985) Manipulation of the morphogenetic pathways of tobacco

explants by oligosaccharins. *Nature (London)*, **314**, 615–617.
Wareing, P.F. (1987) Juvenility and cell determination. In *Manipulation of Flowering*, ed. J.G. Atherton, Butterworth, London, 83–92.
Wolpert, L. (1981) Positional information and pattern formation. *Phil. Trans. Roy. Soc. London*, Ser. B, **295**, 441–450.

Further reading
Bernier, G. (1986) The flowering process as an example of plastic development. In *Plasticity in Plants*, SEB Symposium Vol. 40, eds. D.H. Jennings and A.J. Trewavas, Company of Biologists, Cambridge, 257–286.
Naylor, A.W. (1984) Functions of hormones at the organ level of organization. In *Encyclopedia of Plant Physiology, New Series*, Vol. 10, *Hormonal Regulation of Development II*, ed. T.K. Scott, Springer, Berlin, 172–218.

Chapter 6
Akazawa, T. and Miyata, S. (1982) Biosynthesis and secretion of α-amylase and other hydrolases in germinating cereal seeds. *Essays in Biochemistry*, **18**, 41–78.
Baulcombe, D.C., Martienssen, R.A., Huttly, A.M., Barker, R.F. and Lazarus, C.M. (1986) Hormonal and developmental control of gene expression in wheat. *Phil. Trans. Roy. Soc. London*, Ser. B, **314**, 441–451.
Gilmour, S.J. and MacMillan, J. (1984) Effects of inhibitors of gibberellin synthesis on the induction of α-amylase in embryoless caryopses of *Hordeum vulgare* cv. Himalaya. *Planta (Berlin)*, **162**, 89–90.
Jacobsen, J.V. and Beach, L.R. (1985) Control of transcription of α-amylase and rRNA genes in barley aleurone protoplasts by gibberellin and abscisic acid. *Nature (London)*, **316**, 275–277.
Karssen, C.M., Groot, S.P.C. and Koorneef, M. (1987) Hormone mutants and seed dormancy in *Arabidopsis* and tomato. In *Developmental Mutants in Higher Plants*, SEB Seminar Series 32, eds. H. Thomas and D. Grierson, Cambridge University Press, Cambridge, 119–134.
Quatrano, R.S. (1986) Regulation of gene expression by abscisic acid during angiosperm embryo development. In *Oxford Surveys of Plant Molecular and Cell Biology*, **3**, ed. B.J. Miflin, Oxford University Press, London, 467–477.
Sponsel, V.M. (1985) Gibberellins in *Pisum sativum* — their nature, distribution and involvement in growth and development of the plant. *Physiol. Plant.* **65**, 533–538.
Van Staden, J. (1983) Seeds and cytokinins. *Physiol. Plant.* **58**, 340–346.

Further reading
Bewley, J.D. and Black, M. (1978) *Physiology and Biochemistry of Seeds in Relation to Germination. 1. Development, Germination and Growth*. Springer, Berlin.
Bewley, J.D. and Black, M. (1982) *Physiology and Biochemistry of Seeds in Relation to Germination. 2. Viability, Dormancy and Environmental Control*. Springer, Berlin.
Mayer, A.M. and Poljakoff-Mayber, A. (1982) *The Germination of Seeds*, 3rd edn., Pergamon, Oxford.

Chapter 7
Bandurski, R.S., Schulze, A. and Reinecke, D.M. (1986) Biosynthetic and metabolic aspects of auxins. In *Plant Growth Substances 1985*, ed. M. Bopp, Springer, Berlin, 83–91.
Bouchet, M.-H., Prat, R. and Goldberg, R. (1983) Kinetics of IAA-induced growth in hypocotyl sections of *Vigna radiata*. *Physiol. Plant.* **57**, 95–100.
Brown, B.T., Foster, C., Phillips, J.N. and Rattigan, B.M. (1979) The indirect role of 2,4D in the maintenance of apical dominance in decapitated sunflower seedlings (*Helianthus annuus* L.). *Planta (Berlin)*, **146**, 475–480.

Cleland, R.E. (1985) The role of hormones in wall loosening and plant growth. *Austr. J. Plant Physiol.* **13**, 93–103.

Cosgrove, D. (1986) Biophysical analysis of plant cell growth. *Ann. Rev. Plant Physiol.* **37**, 377–405.

Eisinger, W. (1983) Regulation of pea internode expansion by ethylene. *Ann. Rev. Plant Physiol.* **34**, 225–240.

Fry, S.C. (1986) Cross-linking of matrix polymers in the growing cell walls of angiosperms. *Ann. Rev. Plant Physiol.* **37**, 165–186.

Hanson, J. and Trewavas, A.J. (1982) Regulation of plant cell growth; the changing perspective. *New Phytol.* **90**, 1–21.

Kutschera, U. and Schopfer, P. (1985) Evidence against the acid-growth theory of auxin action. *Planta (Berlin),* **163**, 483–493.

Mertens, R. and Weiler, E.W. (1983) Kinetic studies of the redistribution of endogenous growth regulators in gravi-reacting organs. *Planta (Berlin),* **158**, 339–348.

Phinney, B.O. (1984) Gibberellin A_1, dwarfism and the control of shoot elongation in higher plants. In *The Biosynthesis and Metabolism of Plant Hormones,* SEB Seminar Series 23, eds. A. Crozier and J.R. Hillman, Cambridge University Press, Cambridge, 17–41.

Phinney, B.O. and Spray, C. (1982) Chemical genetics and the gibberellin pathway in *Zea mays* L. In *Plant Growth Substances 1982,* ed. P.F. Wareing, Academic Press, London and New York, 101–110.

Roberts, J.A. (1987) Mutants and gravitropism. In *Developmental Mutants in Higher Plants,* SEB Seminar Series 32, eds. H. Thomas and D. Grierson, Cambridge University Press, Cambridge, 135–154.

Stoddart, J.L. (1987) Genetic and hormonal regulation of stature. In *Developmental Mutants in Higher Plants,* SEB Seminar Series 32, eds. H. Thomas and D. Grierson, Cambridge University Press, Cambridge, 155–180.

Shaw, S. and Wilkins, M.B. (1973) The source and lateral transport of growth inhibitors in geotropically stimulated roots of *Zea mays* and *Pisum sativum. Planta (Berlin),* **109**, 11–26.

Smart, C.C. and Trewavas, A.J. (1983) Abscisic acid-induced turion formation in *Spirodela polyrrhiza* L. I. Production and development of the turion. *Plant Cell Environm.* **6**, 507–514.

Taiz, L. (1984) Plant cell expansion: regulation of cell wall mechanical properties. *Ann. Rev. Plant Physiol.* **35**. 585–657.

Theologis, A. (1986) Rapid gene regulation by auxin. *Ann. Rev. Plant Physiol.* **37**, 407–438.

Vanderhoef, L.N. and Dute, R.R. (1981) Auxin-regulated wall loosening and sustained growth in elongation. *Plant Physiol.* **67**, 146–149.

Wright, M. (1981) Reversal of the polarity of IAA transport in the leaf sheath base of *Echinochloa colonum. J. exp. Bot.* **32**, 159–169.

Further reading

Brummel, D.A. and Hall, J.L. (1987) Rapid cellular responses to auxin and the regulation of growth. *Plant Cell Environm.* **10**, 523–543.

Dennison, D.S. (1984) Phototropism. In *Advanced Plant Physiology,* ed. M.B. Wilkins, Pitman UK, London, 149–162.

Hillman, J.R. (1984) Apical dominance. In *Advanced Plant Physiology,* ed. M.B. Wilkins, Pitman UK, London, 127–148.

Chapter 8

Archbold, D.D. and Dennis, F.G. (1985) Strawberry receptacle growth and endogenous IAA content as affected by growth regulator application and achene removal. *J. Amer. Soc. Hort. Sci.* **110**, 816–820.

Bradford, K.J. and Yang, S.F. (1980) Xylem transport of 1-aminocyclopropane-1-carboxylic acid, an ethylene precursor, in waterlogged tomato plants, *Plant Physiol.* **65**, 322–326.

Brady, C.J. (1987) Fruit ripening. *Ann. Rev. Plant Physiol.* **38**, 155–178.

Brecht, J.K. (1987) Locular gel formation in developing tomato fruit and the initiation of ethylene production. *Hortscience,* **22**, 476–479.

Davies, W.J., Metcalfe, J.C., Schurr, U., Taylor, G. and Zhang, J. (1987) Hormones as chemical signals involved in root to shoot communication of effects of changes in the soil environment. In *Hormone Action in Plant Development — a Critical Appraisal*, eds. G.V. Hoad, J.R. Lenton, M.B. Jackson and R.K. Atkin, Butterworth, London, 201–216.

De Silva, D.L.R., Cox, R.C., Hetherington, A.M. and Mansfield, T.A. (1985) Suggested involvement of calcium and calmodulin in the responses of stomata to abscisic acid. *New Phytol.* **101**, 555–563.

De Silva, D.L.R., Cox, R.C., Hetherington, A.M. and Mansfield, T.A. (1986) The role of abscisic acid and calcium in determining the behaviour of adaxial and abaxial stomata. *New Phytol.* **104**, 41–51.

Grierson, D. (1985) Gene expression in ripening tomato fruit. *CRC Critical Reviews in Plant Sciences*, **3**, 113–132.

Grierson, D., Maunders, M.J., Slater, A., Ray, J., Bird, C.R., Schuch, W., Holdsworth, M.J., Tucker, G.A. and Knapp, J.E. (1986) Gene expression during tomato ripening. *Phil. Trans. Roy. Soc. London*, Ser. B, **314**, 399–410.

Kelly, P., Trewavas, A.J., Lewis, L.N., Durbin, M.L. and Sexton, R. (1987) Translatable mRNA changes in ethylene induced abscission zones of *Phaseolus vulgaris* (Red Kidney). *Plant Cell Environm.* **10**, 11–16.

Klee, H.J., Horsch, R.B., Hinchee, M.A., Hein, M.B. and Hoffmann, N.L. (1987) The effects of overproduction of two *Agrobacterium tumefaciens* T-DNA auxin biosynthesis gene products in transgenic petunia plants. *Genes & Devel.* **1**, 86–96.

Roberts, J.A., Tucker, G.A. and Maunders, M.J. (1985) Ethylene and foliar senescence. In *Ethylene and Plant Development*, eds. J.A. Roberts and G.A. Tucker, Butterworth, London, 267–275.

Roberts, J.A., Grierson, D. and Tucker, G.A. (1987) Genetic variants as aids to examine the significance of ethylene in development. In *Hormone Action in Plant Development — a Critical Appraisal*, eds. G. Hoad, J.R. Lenton, M.B. Jackson and R.K. Atkin, Butterworth, London, 107–118.

Sexton, R., Lewis , L.N., Trewavas, A.J. and Kelly, P. (1985) Ethylene and abscission. In *Ethylene and Plant Development*, eds. J.A. Roberts and G.A. Tucker, Butterworth, London, 173–196.

Sexton, R. and Roberts, J.A. (1982) Cell biology of abscission. *Ann. Rev. Plant Physiol.* **33**, 133–162.

Sexton, R. and Woolhouse, H.W. (1984) Senescence and abscission. In *Advanced Plant Physiology*, ed. M.B. Wilkins, Pitman UK, London, 469–497.

Stead, A.D. (1985) The relationship between pollination, ethylene production and flower senescence. In *Ethylene and Plant Development*, eds. J.A. Roberts and G.A. Tucker, Butterworth, London, 71–81.

Thimann, K.V. (1985) The senescence of detached leaves of *Tropaeolum*. *Plant Physiol.* **79**, 1107–1110.

Tucker, M.L., Christoffersen, R.E., Woll, L. and Laties, G.G. (1985) Induction of cellulase by ethylene in avocado fruit. In *Ethylene and Plant Development*, eds. J.A. Roberts and G.A. Tucker, Butterworth, London, 163–171.

Further reading

Thomas, H. and Grierson, D. (eds.) (1987) *Developmental Mutants in Higher Plants*, SEB Seminar Series 32, Cambridge University Press, Cambridge.

Chapter 9

Brinegar, A.C., Stevens, A. and Fox, J.E. (1985) Biosynthesis and degradation of a wheat embryo cytokinin-binding protein during embryogenesis and germination. *Plant Physiol.* **79**, 706–710.

Evans, D.E., Dodds, J.H., Lloyd, P.C., ap Gwynn, I. and Hall, M.A. (1982) A study of the subcellular localisation of an ethylene binding site in developing cotyledons of *Phaseolus vulgaris. Planta (Berlin),* **54**, 48–52.

Gronemeyer, H. (1985) Photoaffinity labelling of steroid hormone binding sites. *Trends Biochem. Sci.* **10**, 264–267.

Hall, M.A. (1986) Ethylene receptors. In *Hormones, Receptors and Cellular Interactions in Plants,* eds. C.M. Chadwick and D.R. Garrod, Cambridge University Press, Cambridge, 69–89.

Hornberg, C. and Weiler, E.W. (1984) High affinity binding sites for abscisic acid on the plasmalemma of *Vicia faba* guard cells. *Nature (London),* **319**, 321–324.

Jacobs, M. and Short, T.W. (1986) Further characterization of the presumptive auxin transport carrier using monoclonal antibodies. In *Plant Growth Substances 1985,* ed. M. Bopp, Springer, Berlin, 219–226.

Key, J.L., Kroner, P., Walker, J., Hong, J.C., Ulrich, T.H., Ainley, W.M., Gantt, J.S. and Nagao, R.T. (1986) Auxin-regulated gene expression. *Phil. Trans. Roy. Soc. London,* **314**, 427–440.

Libbenga, K.R., Maan, A.C., Van der Linde, P.C.G. and Mennes, A.M. (1986) Auxin receptors. In *Hormones, Receptors and Cellular Interactions in Plants,* eds. C.M. Chadwick and D.R. Garrod, Cambridge University Press, Cambridge, 1–68.

Pauls, K.P., Chambers, J.A., Dumbroff, E.B. and Thompson, J.E. (1982) Perturbation of phospholipid membranes by gibberellins. *New Phytol.* **91**, 1–17.

Ruoho, A.E., Rashidbaigi, A. and Raeder, P.E. (1984) Approaches to the identification of receptors utilising photoaffinity labelling. In *Membranes, Detergents, and Receptor Solubilization,* eds. C.J. Venter and L.C. Hamilton, Alan R. Liss, New York, 119–160.

Stoddart, J.L. (1986) Gibberellin receptors. In *Hormones, Receptors and Cellular Interactions in Plants,* eds. C.M. Chadwick and D.R. Garrod, Cambridge University Press, Cambridge, 91–114.

Thompson, M., Krull, U.J. and Venis, M.A. (1983) A chemoreceptive bilayer lipid membrane based on an auxin-receptor ATPase electrogenic pump. *Biochem. Biophys. Res. Comm.* **110**, 300–304.

Venis, M.A. (1985) *Hormone Binding in Plants.* Longman, New York and London.

Further reading

Venis, M.A. (1987) Hormone receptor sites and the study of plant development. In *Hormone Action in Plant Development — a Critical Appraisal,* eds. G.V. Hoad, J.R. Lenton, M.B. Jackson and R.K. Atkin, Butterworth, London, 53–61.

Chapter 10

Baulcombe, D.C. (1987) Do plant hormones regulate gene expression during development? In *Hormone Action in Plant Development — a Critical Appraisal,* eds. G.V. Hoad, J.R. Lenton, M.B. Jackson and R.K. Atkin, Butterworth, London, 63–70.

Berridge, M.J. (1987) Inositol triphosphate and diacylglycerol: two interacting second messengers. *Ann. Rev. Biochem.* **56**, 159–193.

Cohen, P. (1985) The role of protein phosphorylation in the hormonal control of enzyme activity. *Eur. J. Biochem.* **151**, 429–448.

Ettlinger, C. and Lehle, L. (1988) Auxin induces rapid changes in phosphatidylinositol metabolites. *Nature (London)* **331**, 176–178.

Hepler, P.K. and Wayne, R.O. (1985) Calcium and plant development. *Ann. Rev. Plant Physiol.* **36**, 397–439.

Higgins, T.J.V., Jacobsen, J.V. and Zwar, J.A. (1982) Gibberellic acid and abscisic acid modulate protein synthesis and mRNA levels in barley aleurone layers. *Plant Molec. Biol.* **1**, 191–215.

Marre, E. (1979) Fusicoccin: A tool in plant physiology. *Ann. Rev. Plant Physiol.* **30**, 273–288.

Poovaiah, B.W., Reddy, A.S.N. and McFadden, J.J. (1987) Calcium messenger system: Role of protein phosphorylation and inositol bisphospholipids. *Physiol. Plant.* **69**, 569–573.

Roberts, J.A., Grierson, D. and Tucker, G.A. (1987) Genetic variants as aids to examine the significance of ethylene in development. In *Hormone Action in Plant Development — a Critical Appraisal*, eds. G.V. Hoad, J.R. Lenton, M.B. Jackson and R.K. Atkin, Butterworth, London, 107–118.

Schauf, C.L. and Wilson, K.J. (1987) Effects of abscisic acid on K^+ channels in *Vicia faba* guard cell protoplasts. *Biochem. Biophys. Res. Comm.* **145**, 284–290.

Scherer, G.F.E. (1984) H^+-ATPase and auxin-stimulated ATPase in membrane fractions from Zucchini (*Cucurbita pepo*) and Pumpkin (*Cucurbita maxima*) hypocotyls. *Zeitschr. Pflanzenphysiol.* **114**, 233–237.

Sisler, E.C. (1982) Ethylene binding in normal, *rin* and *nor* mutant tomatoes. *J. Plant Growth Reg.* **1**, 219–226.

Theologis, A. (1986) Rapid gene regulation by auxin. *Ann. Rev. Plant Physiol.* **37**, 407–438.

Van der Linde, P.C.G., Maan, A.C., Mennes, A.M. and Libbenga, K.R. (1985) Auxin receptors in tobacco. In *Proc. 16th FEBS meeting, Moscow*, Part C, ed. Y.A. Orchinnikov, UNU Science Press, Moscow, 397–403.

Zwar, J.A. and Hooley, R. (1986) Hormonal regulation of α-amylase gene transcription in wild oat (*Avena fatua* L.) aleurone protoplasts. *Plant Physiol.* **80**, 459–463.

Chapter 11

Dalziel, J. and Lawrence, D.K. (1984) Biochemical and biological effects of kaurene oxidase inhibitors, such as paclobutrazol. In *Biochemical Aspects of Synthetic and Naturally Occurring Plant Growth Regulators*, British Plant Growth Regulator Group, Monograph 11, eds. R. Menhenett and D.K. Lawrence, 43–57.

Fenton, R., Mansfield, T.A. and Jarvis, R.G. (1982) Evaluation of the possibilities for modifying stomatal movement. In *Chemical Manipulation of Crop Growth and Development*, ed. J.S. McLaren, Butterworth, London, 19–37.

Garrod, J.F. (1982) The discovery and development of plant growth regulators. In *Plant Growth Regulator Potential and Practice*, ed. T.H. Thomas, BPCC Publications, London, 29–56.

Hedden, P. and Graebe, J.E. (1985) Inhibition of gibberellin biosynthesis by paclobutrazol in cell-free homogenates of *Cucurbita maxima* endosperm and *Malus pumila* embryos. *J. Plant Growth Reg.* **4**, 111–122.

Lurssen, K. (1987) The use of inhibitors of gibberellin and sterol biosynthesis to probe hormone action. In *Hormone Action in Plant Development — a Critical Appraisal*, eds. G.V. Hoad, J.R. Lenton, M.B. Jackson and R.K. Atkin, Butterworth, London, 133–144.

Lurssen, K. and Konze, J. (1985) Relationship between ethylene production and plant growth after application of ethylene releasing plant growth regulators. In *Ethylene and Plant Development*, eds. J.A. Roberts and G.A. Tucker, Butterworth, London, 363–372.

Morgan, P.W. (1986) Ethylene as an indicator and regulator in the development of field crops. In *Plant Growth Substances 1985*, ed. M. Bopp, Springer, Berlin, 375–379.

Nichols, R. and Frost, C.E. (1985) Post-harvest effects of ethylene on ornamental plants. In *Ethylene and Plant Development*, ed. J.A. Roberts and G.A. Tucker, Butterworth,

London, 343–351.

Quarrie, S.A. and Gale, M.D. (1986) Examples of the use of natural and synthetic plant growth substances in cereal breeding. In *Plant Growth Substances 1985*, ed. M. Bopp, Springer, Berlin, 404–409.

Further reading

Bruinsma, J. (1985) Plant growth regulators, past and present. In *Growth Regulators in Horticulture*, British Plant Growth Regulator Group Monograph 13, eds. R. Menhenett and M.B. Jackson, 1–13.

Nickell, L.G. (1982) *Plant Growth Regulators — Agricultural Uses*. Springer, Berlin.

Thomas, T.H. (1985) Chemical manipulation of standing crops. In *Plant Products and the New Technology*, Annual Proceedings of the Phytochemical Society, Vol. 26, eds. K.W. Fuller and J.R. Gallon. Oxford University Press, Oxford, 73–90.

Index

abscisic acid
  α-amylase activity and   158
  abscission and   131
  apical dominance and   109
  binding sites for   145 – 6
  biosynthesis of   21 – 3
  biosynthetic sites for   52
  bud dormancy and   110
  Ca^{2+} transport and   152
  carriers for   59
  conjugates of   25
  K^+ transport and   152
  leaf growth and   115
  leaf senescence and   126
  metabolism of   23 – 6, 83
  mutants deficient in   22, 23, 26, 84,
    109, 115, 119
  regulation of gene expression for
    158
  root gravitropism and   104
  seed development and   83, 158
  seed dormancy and   84 – 5
  stomatal aperture   118 – 19
  uptake of   59
abscisin II   21
abscission   129 – 32
  and gene expression   132
abscission zone   61, 129
Acer saccharinum
  seed dormancy in   84
acid-growth theory   95 – 6
S-adenosyl methionine   27, 31
adenine, metabolism of   16
affinity labelling   138 – 40
Agrobacterium tumefaciens   8, 18, 19
aleurone layer   61, 88, 90, 91, 155
aleurone protoplasts   91
α-amylase   25, 88, 90, 91, 155, 156, 161
aminocyclopropane carboxylic acid
  (ACC)   27, 120, 171
  transport of   56
aminocyclopropane carboxylic acid
  (ACC) synthase
  after IAA treatment   54

and ripening   121, 123
and wounding   54
isolation and properties of   28 – 9
aminoethoxyvinylglycine (AVG)   24, 53,
  103, 109, 120, 127, 171
aminooxyacetic acid (AOA)   29
AMO 1618   14
amyloplast   100, 102, 103
antisense mRNA   163
antitranspirants   169 – 70
apical dominance
  in roots   109
  in shoots   106 – 9
Arabidopsis thaliana
  mutants of   84, 87
  seed dormancy in   84, 87
auxin   (*see also* indole-3-acetic acid)
  abscission and   130 – 1
  apical dominance and   106
  ATPase activity and   151
  binding sites for   139 – 44
  biosynthesis of   4 – 7, 8
  fruit growth and   116
  efflux of   57, 58
  gravitropism and   100 – 3
  H^+ efflux and   95 – 6, 151
  metabolism of   7 – 8, 82
  phototropism and   105, 106
  regulation of gene expression for   95,
    97, 155
  vascular differentiation and   75
Avena fatua
  IAA analysis in   44
  seed dormancy in   84
A. sativa
  aleurone tissue in   25, 90
  α-amylase in   156 – 7

'bakanae' disease   9
barley aleurone   25, 90
β-cyanoalanine synthase   29
β-galactosidase and growth   98
β-hydroxylation of gibberellins   15, 55,
  81, 145

bioassays 36 – 8
  limits of detection of 37
Brassica
  auxin biosynthesis in 5
  seed development in 83, 158
  vernalization of 72
brassinosteroids 1
bromeliads
  flowering in 71, 170
bud dormancy 110

calcium ion (Ca^{2+})
  gravitropism and 103, 105
  growth and 96
  stomatal aperture and 118
cAMP 161
calmodulin 161
carnation, flowering in 172
CBF-1 148, 149
cell signalling 159 – 63
cell wall, structure of 97
cellulase
  abscission and 131 – 2
  ripening and 124
cerone 170
cerulenin 96
chemical affinity labelling 140
chemical assays for PGRs 38
chemiosmotic hypothesis 56 – 7, 141
Chenopodium amaranticolor, flowering
  in 70
chlormequat (CCC) 14, 166 – 8
4-chloro-3-indole acetic acid 5
Cholodny—Went hypothesis 100, 102,
  103
climacteric fruits 121 – 5
cocklebur
  seed dormancy in 86
Coleus, vascular differentiation in 75
columella 100
Commelina communis, regulation of
  stomatal aperture in 119, 152,
  153, 161
commercial PGRs 164 – 73
controlled atmosphere storage 171
copalylpyrophosphate 10, 12, 167
cotton crop 171
coulometric detector 44
correlative inhibition 106 – 9
criteria, PGR binding 134 – 5
crown gall 8, 18, 19
Cucumis sativus, sex expression in 73
C. maxima, GA biosynthesis in 167
C. pepo, IAA transport in 57, 59

cytokinins
  apical dominance and 108
  binding sites for 146 – 9
  biosynthesis of 15 – 19
  biosynthetic sites for 52 ˙
  bud dormancy and 65, 111
  in caulonemal development in *Funaria*
    160
  conjugation of 20
  hydrolysis of 19
  in seeds 81 – 2
  leaf growth and 115
  leaf senescence and 126
  metabolism of 20, 82
  mutants deficient in 21
  ripening and 124
  seed development and 81 – 2
  stomatal aperture and 118
  transport and water stress and 60
  tuberization and 65 – 6, 112
cytokinin oxidase 20

dextranase 98
diacylglycerol 163
differentiation 68 – 79
Digitalis purpurea, flower senescence
  in 128
dihydrophaseic acid 25, 26, 83
dormancy 65 – 7, 83 – 6
dormin 21, 110
dwarf mutants 11, 14, 100

early hydroxylation pathway 10, 13
Echinochloa colonum
  gravitropism of node of 103
  IAA transport in 60
elastic extensibility 94, 98
electron capture detector 42
embryogeny 83, 84, 158
endodermis 100, 102
ent-kaurene 10, 14, 51, 52, 167
ent-kaurene synthetase 14
enzyme-linked immunoassay
  (ELISA) 39
epinasty 119 – 21
equilibrium dialysis 137
ethephon 170 – 1
ethylene
  abscission and 129 – 30
  apical dominance and 109
  binding sites for 149
  biosynthesis of 26 – 9
  biosynthetic sites for 52, 53
  compounds generating 170 – 1

epinasty and   120
extraction of   35
flower senescence and   128
fruit ripening and   121, 122
H^+ efflux and   96
leaf senescence and   126
metabolism of   29 – 31
regulation of gene expression
   for   158 – 9
wall extensibility and   94
ethylene-forming enzyme   29
   activity of during ripening   122, 123
ethylene glycol   30
ethylene oxide   30, 31
ethylene-suppressing compounds
   171 – 2
explants   131

farnesol   169
flame ionization detector   42
floral inhibitors   70
florigen   70, 71, 78
flower senescence   127
flowering   69 – 73
fluorescence detector   44
fluoroimmunoassay   40
fluridone   23
fruit growth   115 – 6
fruit ripening   121 – 5
Funaria, bud formation in   160
fungicides   168
fusicoccin (FC)   87
   growth and   95, 96
   H^+ efflux and   152
   ion transport and   152, 153
   phosphorylation and   161

GA_{12}-aldehyde   10, 11, 12
gas chromatography   36, 41 – 2
gas chromatography-mass spectrometry
   (GCMS)   39, 42 – 3
gel filtration   137
genetic engineering   173
geolectric effect   103
geranylgeranyl pyrophosphate   10, 12,
   14, 167
germination   86 – 7
Gibberella fujikuroi   9, 10, 166
gibberellic acid   10
gibberellins (GAs)
   α-amylase and   155, 156, 161
   biosynthesis of   8 – 15, 51
   biosynthetic sites for   51 – 2
   binding sites for   144 – 5

bud dormancy and   110
enzyme secretion and   91, 92
flowering and   71
fruit growth and   116
gravitropism and   102
growth and   94
H^+ efflux and   96
hyponasty and   121
leaf growth and   114 – 15
metabolism of   15, 81
mutants deficient in   11, 87, 100, 101,
   102
mutants insensitive to   100
regulation of gene expression
   for   155, 156, 157
ripening and   124
seed development and   47, 80
seed germination and   86
vernalization and   73
xylogenesis and   77
wall extensibility and   94
grass nodes   103
gravitropic mutants   see mutants
gravitropism
   in nodes   102
   in roots   103
   in shoots   100 – 2
growth
   of intact plants   99
   of water plants   96
growth retardants   99, 166

hazel, dormancy of seed of   86
Helianthus annuus
   apical dominance in   107 – 8
   epinasty in   120
   gravitropism in   102
   phototropism in   106
herbicides   165
high performance liquid chromatography
   (HPLC)   36, 44, 47
hormone concept   49 – 50
hydraulic conductance   93, 94
hydrolytic enzymes, abscission and   132
Hyoscyamus niger, flowering in   70
hyponasty   120, 121

illuminating gas   26, 130
immunoaffinity chromatography   39
immunolocalization   48
immunological methods   38 – 41
indole-3-acetaldehyde   5
indole-3-acetaldoxime   5
indole-3-acetamide   8

indole-3-acetic acid (IAA)
  abscission and 130−1
  apical dominance and 107−9
  ATPase activity stimulation by 96
  binding sites for 139−44
  biosynthesis of 4−7, 8
  conjugation of 7, 82
  decarboxylation of 7, 8, 9
  epinasty and 120
  gravitropism and 100−3
  H⁺ efflux and 95−6, 151
  leaf growth and 114
  myoinositol conjugate of 7, 51, 82
  oxidation of 8, 9
  phototropism and 105−6
  regulation of gene expression for 95, 97
  transport of 55, 56, 59, 60
  wall extensibility and 94, 95
indole-3-acetic acid transport
  effect of ageing on 60
  effect of ethylene on 60
  effect of gravity on 60
  in-vitro techniques for assessing 59
  in roots 56
  in shoots 55, 56
indole-3-acetonitrile 5
indole-3-aldehyde 7, 9
indole-3-butyric acid 165
indole-3-methanol 7
indole-3-pyruvic acid 5
induction, vascular 75
inhibitor β 21
inhibitors, GA-biosynthesis 99, 166−8
inositol phospholipids 163
inositol 1, 4, 5-triphosphate 163
internal standards for PGR analysis 35, 36
ion movements, PGRs and 151
Ipomea tricolor, ethylene metabolism in 30
ivy, phase transitions in 68

jasmonic acid 126
juvenility 68, 69

kinetin 16, 17

Lactuca sativa 84
  seed dormancy in 84
leaves
  abscission of 25, 129
  growth of 114, 115
  senescence of 126, 127

light, root gravitropism and 104
localization of auxin binding sites 143
long-day plants (LDP) 69, 70, 71

malonyl-ACC 27
mass spectrometer detectors 44
methionine 27
3-methylene oxindole 7, 9
methylthioadenosine 27
methylthioriboside 27
mevalonic acid 10, 12
mobilization of storage reserves 88−92
monocarpic senescence 129
monoclonal antibodies 39
morphogenesis 77−9
mutants
  ABA-deficient 22, 23, 26, 84, 109, 115, 119
  CK-deficient 21
  developmental 133
  dwarf 11, 14, 100
  GA-deficient 11, 87, 100, 101, 102
  GA-insensitive 100
  gravitropic 105
  in tomato 22, 23, 124, 159
naphthylphthalamic acid (NPA)
  and auxin efflux 58
  binding sites for 58, 139, 141
  inhibition of auxin transport by 56
Nicotiana tabacum
  auxin binding in 143, 161
  CK biosynthesis in 17
  IAA biosynthesis in 5
non-climacteric fruit 121, 122
norflurazon 23
nutrient diversion hypothesis 109
Nymphoides, ethylene-promoted growth in 96

oligosaccharides, flowering and 71
osmosis, cell growth and 93
oxindole-3-acetic acid 8, 9

paclobutrazol (PP333) 166, 167, 168
parthenocarpy 116
pattern formation 74
peach, seed dormancy in 84
pea *see Pisum*
Pharbitis nil, flowering in 70
phase changes 69
phaseic acid 23, 25, 26, 83
P. coccineus
  ABA transport in 59
  GAs in seeds of 15

P. vulgaris
  apical diminance in   107
  ethylene binding in   149
  leaf growth in   114
  RNAs during abscission in   132
phenylacetic acid   5
photoaffinity labelling   139, 140, 145
photoionization detector   42
photoperiodism   69
phototropism
  in coleoptiles   4, 49, 105, 106
  in green plants   105, 106
Physcomitrella patens, CK mutant of
  21
Pinus sylvestris, IAA metabolism in   7
Pisum
  ethylene metabolism in   29
  GAs in seeds of   51, 81
  growth in   94
  triple response in   31
plant growth regulators (PGRs)   1
  binding studies with   134 – 50
  biosynthesis and metabolism of
    4 – 32
  extraction of   34
  gene expression for   153 – 9
  purification of   35 – 6
  as second messengers   159 – 63
  sensitivity to   49, 62 – 7
  transport of   55 – 61
plant growth substances   1
plastic extensibility   94, 98
polyamines   1
  biosynthesis of   31
  metabolism of   32
  senescence and   127
polygalacturonase
  abscission and   131, 132
  ripening and   124, 159
positional differentiation   77
positional information   77 – 8
potato
  tuber growth in   65
  tuber dormancy in   66 – 7
protein kinases   161
putrescine   31, 32
pyridoxalphosphate-mediated enzymes
  29

radioimmunoassay (RIA)   39
reflection coefficient   93
rhizobitoxin and ethylene production
  171
ripening   121 – 5

root growth   96
rooting compounds   165

Scatchard plot   136
second messengers   152, 159 – 63
seed development   80 – 3
seed dormancy   83 – 6
selective ion monitoring   44
senescence   125 – 9
separation zone   129
sex expression   73 – 4
shoot growth   93 – 100
short-day plants (SDP)   69, 70
Silene armeria, flowering in   71
silver thiosulphate and flower longevity   172
soluble binding sites   143
spectrofluorimetry   38
spermidine   31
spermine   31
spinach, sex expression in   73
Spirodella polyrrhiza, turion induction in
  110, 111, 112
statocytes   100
statoliths   100
stomatal aperture   117
  (see also Commelina communis)
stomatal sensitivity   119
sterol biosynthesis   168
synthetic auxins   165

Taraxacum officinale, regulation of
  senescence in   126
target cells   61
tomato, mutants of
  flacca   22, 23
  notabilis   22, 23
  rin   124, 159
  sitiens   22, 23
transgenic plants   33, 133, 174
transport of IAA   55 – 60
transposable elements   14
transposon tagging   14
triiodobenzoic acid (TIBA)   56, 58, 107,
  141, 165, 166
  and auxin efflux   58
triple response   26, 31, 37
tropisms   100 – 6
tryptamine   5
tryptophan aminotransferase   5
tryptophan, IAA biosynthesis and
  5 – 7, 8
tryptophan monooxygenase   8
tuberization   112
turion induction   110, 111, 112

vanadate 96, 99
vascular differentiation 75
vernalin 72, 78
vernalization 72, 73
Vicia faba
  CK binding in 52
  CK metabolism in 20
violaxanthin 21
vivipary 83

wall extensibility 93, 94, 97, 115
wall loosening factor 95
water stress, ABA and 52
wounding, vascular differentiation and 75

Xanthium, leaf senescence in 126

xylem, differentiation of 75 – 6
xylogenesis 75 – 6

yield threshold 94

Zea mays
  auxin binding in 141, 142
  GA mutants of 11, 14, 100, 101
  gravitropism in 102, 104
  IAA conjugation in 51
  IAA metabolism in 8
  seed cytokinins of 81
  sex expressions in 73
zeatin 16
Zinnia elegans, xylogenesis in 75
zucchini, *in-vitro* IAA transport in 59